Psychological Governance and Public Policy

There have been significant developments in the state of psychological, neuroscientific and behavioural scientific knowledge relating to the human mind, brain, action and decision-making over the past two decades. These developments have influenced public policy making and popular culture in the UK and elsewhere – through policies and emerging social practices focused on behavioural change, happiness, wellbeing, therapy, resilience and character. Yet little attention has been paid to examining the wider political and ethical significance of the widespread use of psychological governance techniques. There is a pressing and recognised need to address the behaviour change agenda in relation to how our cultural ideas about the brain, mind, behaviour and self are changing.

This book provides a critical account of existing forms of psychological governance in relation to public policy. It asks whether we can speak of a co-ordinated and novel shift in governance or, rather, whether these trends are more simply pragmatic policy tools based on advances in scientific evidence. With contributions from leading scholars across the social sciences from the UK, the USA and Canada, chapters identify practical, political and research challenges posed by the current policy enthusiasm for particular branches of affective neuroscience, behavioural economics, positive psychology and happiness economics. The core focus of this book is to investigate the ways in which knowledge about the mind, brain and behaviour has informed the methods and techniques of governance and to explore the implications of this for shaping citizen identity and social practice.

This groundbreaking book will be of interest to students, scholars and policy-makers interested in and working within geography, economics, sociology, psychology, politics and cultural studies.

Jessica Pykett is a social and political geographer at the University of Birmingham, UK.

Rhys Jones is Professor of Human Geography at Aberystwyth University, UK.

Mark Whitehead is Professor of Human Geography at Aberystwyth University, UK.

Routledge Research in Place, Space and Politics

Series Edited by Professor Clive Barnett, Professor of Geography and Social Theory, University of Exeter, UK.

This series offers a forum for original and innovative research that explores the changing geographies of political life. The series engages with a series of key debates about innovative political forms and addresses key concepts of political analysis such as scale, territory and public space. It brings into focus emerging interdisciplinary conversations about the spaces through which power is exercised, legitimized and contested. Titles within the series range from empirical investigations to theoretical engagements and authors comprise of scholars working in overlapping fields including political geography, political theory, development studies, political sociology, international relations and urban politics.

Psychological Governance and Public Policy

Governing the mind, brain and behaviour

Edited by
**Jessica Pykett, Rhys Jones and
Mark Whitehead**

Routledge
Taylor & Francis Group

LONDON AND NEW YORK

First published 2017
by Routledge
2 Park Square, Milton Park, Abingdon, Oxon OX14 4RN

and by Routledge
711 Third Avenue, New York, NY 10017

First issued in paperback 2018

Routledge is an imprint of the Taylor & Francis Group, an informa business

British Library Cataloguing in Publication Data
A catalogue record for this book is available from the British Library

Library of Congress Cataloging in Publication Data
Names: Pykett, Jessica, editor. | Jones, Rhys, 1971- editor. | Whitehead, Mark, 1975- editor.
Title: Psychological governance and public policy : governing the mind, brain and behaviour / edited by Jessica Pykett, Rhys Jones and Mark Whitehead.
Description: Abingdon, Oxon ; New York, NY : Routledge, 2017. | Series: Routledge research in place, space and politics series | Includes bibliographical references and index.
Identifiers: LCCN 2016020413 | ISBN 9781138930735 (hardback) | ISBN 9781315680248 (ebook)
Subjects: LCSH: Political psychology. | Political participation--Psychological aspects. | Social psychology--Political aspects.
Classification: LCC JA74.5 .P76 2017 | DDC 320.01/9--dc23
LC record available at https://lccn.loc.gov/2016020413

ISBN 13: 978-1-138-62421-4 (pbk)
ISBN 13: 978-1-138-93073-5 (hbk)

Typeset in Times New Roman
by Taylor & Francis Books

Contents

Acknowledgement

We are grateful to all the participants and speakers involved in the ESRC Seminar Series on Behaviour Change and Psychological Governance (2013–2015), who have shaped this book, and to the ESRC for funding this series (Grant number: ES/L000296/1). The seminars were held at the Universities of Bristol, Birmingham, Durham and Aberystwyth and at the Royal Society for the Arts (RSA), London. Speakers providing insight on neuroarchitecture, environmental psychology and urban design included: Monica Degen, Eve Edelstein, Graeme Evans, Dan Lockton and Margaret Tarampi. Speakers on the complex theoretical entanglements of 'the psychological' with various objects, habits and forces included: Lisa Baraitser, David Bissell, Lisa Blackman, Felicity Callard, Des Fitzgerald and Graham Harman. Speakers on the ethical dilemmas of applying behavioural insights in policy included: Leigh Caldwell, David Chandler, Steven Johnson, Adam Oliver, Luke Perry and Rory Sutherland. Speakers at the Aberystwyth Summer School on Critical Dialogues on Psychology, Behaviour and Brain Science included: Rupert Alcock, Diana Beljaars, Felicity Callard, Osian Elias, Rachel Lilley, Charlotte Longden, Steven Stanley, Helen Stinson and Roger Tyers. Speakers on the politics and economics of attention were: Clive Barnett, Matthew Crawford, Peter Doran, Matt Hannah and Sam Kinsley. Finally, we of course thank the contributors to this book and the additional speakers at the seminar on psychological resilience on which this book is largely based, including: Erica Burman, Jan de Vos and Vanessa King along with Will Leggett who assisted in chairing the seminar. We would like to thank our co-organisers of the seminar series, Ben Anderson, J.-D. Dewsbury, Maria Fannin and Joe Painter as well as Steven Johnson who collaborated on the RSA seminar. We are grateful to Stacey Smith, Colin Lorne and Rupert Alcock who provided commentaries on the seminars.

Contributors

Sam Binkley is Associate Professor of Sociology at Emerson College, Boston, MA., USA. He has published articles on the historical and social production of subjectivity in varied contexts, chiefly through theoretical engagement with the work of Michel Foucault, and has an empirical interest in popular psychology. He is a co-editor of *Foucault Studies* and author of *Getting Loose: Lifestyle Consumption in the 1970s* (Duke University Press, 2007) and *Happiness as Enterprise: An Essay on Neoliberal Life* (SUNY, 2014). His current research considers the wider problematic of anti-racism, understood as governmental imperative. His research is available at: http://sambinkley.net.

Suparna Choudhury, PhD is an Assistant Professor in the Division of Social and Transcultural Psychiatry at McGill University, Montreal. She works on the social and clinical contexts of research on the adolescent brain and, more broadly, on the interdisciplinary programme of critical neuroscience. She did her doctoral research in cognitive neuroscience at University College London and her postdoctoral research in transcultural psychiatry at McGill University. Prior to her current position, she led a research group on Constructions of the Brain at the Max Planck Institute for History of Science, Berlin.

William Davies is Senior Lecturer at Goldsmiths, University of London, UK., where he is Co-Director of the Political Economy Research Centre. He is author of *The Limits of Neoliberalism: Authority, Sovereignty and the Logic of Competition* (Sage, 2014) and *The Happiness Industry: How the Government and Big Business Sold Us Wellbeing* (Verso, 2015).

Kathryn Ecclestone is Professor of Education at the University of Sheffield, UK. Her research explores the political and cultural rise of 'therapeutic culture' in a growing number of countries, reflected in the therapisation of policy and practice around interventions for 'emotional well-being' and 'resilience' across social policy, including education and family interventions. She is particularly interested in the ways in which the growth of behavioural interventions in educational settings, together with formal and informal assessments of people's emotional capabilities, reflect and encourage therapisation. Her books include *Emotional Well-Being in Policy and Practice: Interdisciplinary Perspectives* (Routledge, 2013) and *The Dangerous Rise of Therapeutic Education* (Routledge, 2008 with Dennis Hayes).

Rosalind Edwards is Professor of Sociology and Social Sciences Director of Research and Enterprise at the University of Southampton, UK. She is a co-director of the ESRC National Centre for Research Methods and an elected member of the Academy of Social Sciences. She has researched and published widely in the areas of family life and policies and is founding and co-editor of the *International Journal of Social Research Methodology.*

Val Gillies is a Professor of Social Policy at the University of Westminster and has researched and published widely in the area of family, social class and marginalised children and young people, producing an extensive range of journal articles, books and chapters on parenting, youth, behaviour support policies in schools and home school relations as well as qualitative research methods. Her latest book *Pushed to the Edge: Inclusion and Behaviour Management in Schools* (Policy Press, 2016) is based on a long-term ethnography of young people referred to in-school behaviour support units. She has also recently completed an ESRC-funded historical comparative analysis of policy and practice constructions of 'troubled families'.

Peter John is Professor of Political Science and Public Policy in the School of Public Policy, University College London, UK. He is known for his books on public policy: *Analysing Public Policy* (Routledge, 2012, second edition) and *Making Policy Work* (Routledge, 2011). He is author, with Keith Dowding, of *Exits, Voices and Social Investment: Citizens' Reaction to Public Services* (Cambridge University Press, 2012) and, with Anthony Bertelli (NYU), *Public Policy Investment* (Oxford University Press, 2013). He uses experiments to study civic participation in public policy with the aim of finding out what governments and other public agencies can do to encourage citizens to carry out acts of collective benefit. This work appeared in *Nudge, Nudge, Think, Think: Using Experiments to Change Civic Behaviour* (Bloomsbury Academic, 2011). He is an academic advisor to the Behavioural Insights Team and is involved in a number of projects that seek to test out behavioural insights with trials, such as the redesign of tax reminders and channel shift. He co-edits *The Journal of Public Policy* for Cambridge University Press.

Rhys Jones is Professor of Human Geography at Aberystwyth University, Wales. Rhys' work lies at the intersection between political, historical and cultural geography and focuses in particular on the various geographies of the state and its related group identities. He has addressed the geographies of the state in a variety of contexts, ranging from the various organisational, territorial and cultural changes associated with the state-making process in medieval Wales to the more contemporary processes of territorial and functional restructuring affecting the UK state. His work on these themes has appeared in a variety of different journal articles as well as a monograph – *People/States/Territories* – published in 2007 by Blackwell as part of the RGS-IBG Book Series.

The Midlands Psychology Group (John Cromby, Bob Diamond, Paul Kelly, Paul Moloney, Penny Priest, Jan Soffe-Caswell) was founded by David Smail (1938–2014) in 2002. They are a close-knit group of clinical, counselling and academic psychologists who write collectively about issues related to psychology, mental health and psychological therapy. They publish journal articles and special issues, organise and speak at conferences, and are currently writing their first book. They can be contacted at: admin@midpsy.org

Jessica Pykett is a social and political geographer at the University of Birmingham, UK. Her research to date has focused on the geographies of citizenship, education, behavioural forms of governance and the influence of applied and popular neuroscience on policy and practice. She teaches on the spatial politics of welfare, work and wealth. Her books include *Re-educating Citizen: Governing Through Pedagogy* (Routledge, 2012), *Changing Behaviours: On the Rise of the Psychological State* (Edward Elgar, 2013 with Rhys Jones and Mark Whitehead) and *Brain Culture: Shaping Policy through Neuroscience* (Policy Press, 2015).

Alberto Sánchez-Allred, PhD teaches in the Humanities, Philosophy, and Religion Department at John Abbott College, Quebec. He did his doctoral research in psychological anthropology at the University of California, Berkeley, and held a postdoctoral position at McGill University in transcultural psychiatry while co-ordinating the Multicultural Mental Health Resource Centre at the Jewish General Hospital in Montreal. His current research is located at the intersections of philosophy and therapeutic practice, old and new.

Mark Whitehead is Professor of Human Geography at Aberystwyth University, Wales. His work focuses on various aspects of environmental governance. His most recent research has considered the emerging impacts of the behavioural sciences on the design and implementation of public policy.

1 Introduction

Psychological governance and public policy

Jessica Pykett, Rhys Jones and Mark Whitehead

The past two decades have seen concerted shifts in the rationales, techniques and methods of public policy making and governance, which have been well documented in social and political science (Rhodes, 1997; Newman, 2001; Bovaird and Löffler, 2003; Le Grand, 2003). Since at least the late 1990s, there has been an increased policy emphasis on enhancing citizen involvement in government through personalising responsibility, tailoring public services to citizen-consumers and co-producing policy in dialogue with representative communities. These changes have been particularly marked in the UK. Yet conversely, we have also witnessed a move away from the traditional channels of representative parliamentary democracy towards the increasing dominance of expert- and evidence-based policy, focusing on 'what works' – a trend prevalent in both the UK and the USA (Sanderson, 2002). This has included more experimental forms of policy trialling, development and adaptation that are informed by 'design thinking' (Bason, 2014) and 'nudging' people towards making decisions in their own best interests by shaping the environments in which decisions are made and clearing the psychological ground for more rational behaviours (Thaler and Sunstein, 2008). This move is inspired by a perceived need to innovate, to provide creative, future-proof solutions and to adopt policies shaped around the needs, aptitudes and indeed the technological and behavioural habits of service 'users'. Such design thinking has been prominent in countries such as Denmark and Singapore and, more recently, in the UK where the Government Office for Science Foresight team, the Cabinet Office Policy Lab and the Behavioural Insights Team (BIT) have played key roles.

There is now also a sense within the policy-making process that pragmatic, efficient and cost-effective policy change can and should be delivered through new forms of discursive fora and co-produced through participatory engagement with citizens (Mahony, 2010). This can involve, for instance, getting the best experts in a room together and 'workshopping' ideas, rapid prototyping, agile development, experimentation, trials, local pilot projects and rolling out change through government innovation networks, perhaps communicated through stylish infographics and facilitated by market research companies, social marketers and communications agencies. Crucially, these new forms of public policy making represent citizens' needs, values, attitudes, preferences and behaviours to policy-makers through specifically mediated channels, such as public opinion polling, focus group research or community consultation initiatives. Sometimes those mediators are academic researchers, perhaps giving evidence to parliamentary or presidential committees, conferences or proceedings. They might also be more self-organising groups such as political lobbyists, pressure groups, advocacy organisations or initiators of online petitions. But increasingly, it is a cadre of *behavioural* experts and consultants operating within the commercial or social enterprise sphere who are called upon to provide policy advice and contribute to policy strategy, design, testing and implementation. In all these cases, considerable work goes into constructing authoritative claims to knowing how and why citizens behave in certain ways and how their behaviour can be changed in the course of addressing specific policy problems.

In this context, the 'behaviour change industry' has emerged as a body of actors – sometimes governmental, sometimes commercial, sometimes third sector organisations (and often a mixture of these) – who are skilled in identifying, delimiting, measuring, modelling, changing and evaluating the behaviour of individual citizens, communities or particular social groups. In particular, this industry draws on a medical paradigm (e.g. randomised controlled trials – see John, this volume) and the theoretical precepts and experimental methods associated with psychology, behavioural economics and neuroscience as both rationale for and means of achieving specific public policy goals. This behaviour change industry has only recently grown in global significance, playing a crucial role in shaping psychological forms of governance. The notion of an emergent *industry* denotes the work and effort that has been involved in the construction of contemporary formations of psychological governance.

This book considers the research, policy and practical challenges associated with psychological governance where behavioural change is posed as a means and an end of liberal governance. We define psychological governance as forms of largely state-orchestrated public policy activity (though 'non-state' actors are widely involved) that aim to shape the behaviour of individuals, social groups or whole populations through the deployment of the insights of behavioural and psychological sciences. The book considers the varied scope and scale of psychological governance techniques and examines to what extent we can talk of a co-ordinated shift in governance as opposed to a pragmatic set of techniques for improving the efficacy of policy-making in straightened financial times. Contributing authors provide analytical accounts of the wider political significance of psychological governance by investigating what kinds of knowledge claims are made in support of it, its historical and sociological significance, how it operates and its methodological precepts, and its effects in terms of citizen-subject formation and the framing of social and mental problems.

Psychological governance specifically denotes (public, commercial and/or non-governmental) interventions targeted at the interface of conscious and non-conscious thought and action, connecting emotional response and rational deliberation. Chief Executive of the UK's BIT, David Halpern, has described this interface in the following terms:

Behind the shroud of our consciousness, a myriad processes race to work out what is going on in the world around us, and how we should respond ... our brains ceaselessly infer, overlay and interpret new information and memories. It's an incredible performance.

(2015: 6)

On the one hand, while these cognitive heuristics are impressive basic human functions to be celebrated, it is clear that contemporary forms of psychological governance are focused on our human tendencies to get things wrong, make bad decisions and deviate from the rational economic actor proposed in classical economic theory. Hence, for Halpern,

[t]he limits of human cognitive capacities, and the naivety and failures of classical economic models, create a powerful case for more regulation and a more active state according to some.

(2015: 6)

This means the state should orchestrate regulatory activity around a more complex and messier vision of the cognitive capacities of individuals, involving the redesign of government business around citizens who, in Thaler and Sunstein's (2008) terms, are emotional *humans*, not rational *econs*. The notion of 'state orchestration' does not suggest that a monolithic and authoritarian state is imposing such forms of intervention on an unsuspecting public, but refers to the explicit state support and development of psychological governance by nation states, which we outline in the following section. The 'behaviour change industry' thus refers to a confluence of state and non-state actors and existing initiatives/policies/programmes as well as a more fluid set of political ideas, agendas and discursive practices that have risen in prominence globally since the turn of the twenty-first century.

Applying behavioural insights to public policy: a global agenda

There are several purposes for which national governments are currently mobilising psychological insights in the cultivation of specific behavioural responses amongst individual citizens and national populations. Evidence of behaviourally informed policy initiatives can be found in Australia, Denmark, Singapore, the UK, the USA and the Netherlands, with rumours afoot of moves to establish state-run behavioural insight teams in Germany, Italy and Canada. The UK's BIT established in 2010 has pioneered several public policy initiatives informed by behavioural science and aimed at changing citizen behaviours for the public good (for detailed histories of the BIT, see Halpern, 2015 and Jones *et al.*, 2013). These initiatives are extremely wide-ranging and include: encouraging tax compliance; reducing missed hospital appointments; designing web-based public health campaigns; reducing drop-outs from adult literacy and numeracy skills programmes; reducing mobile phone theft; increasing the likelihood of Army reservists completing the application process; increasing diversity in the police force; encouraging illegal migrants at Home Office reporting centres to voluntarily return home; changing job centre processes to harness job seekers' commitments to find work; and a number of policy experiments in the fields of consumer protection, charitable giving and international development (Behavioural Insights Team, 2015). This by no means exhaustive list could be supplemented by the interventions, governing cultures and practices that have been more indirectly shaped by the enthusiasm of national governments to support behaviour change as a paradigm for public policy reform.

The use of psychological knowledge in shaping conduct is thus by no means the preserve of national governments. Supranational organisations such as the World Bank, the World Economic Forum, the Organisation for Economic Co-operation and Development (OECD) and the European Union have also reported on the potential of behavioural economic and neuroscientific research to inform a wide range of policy areas, whilst international aid and development organisations have arguably long engaged in behaviour change interventions and communications projects in the areas of public health and poverty alleviation. Nor are these behavioural interventions the preserve of state authorities. Advertisers, marketers, political campaigners, NGOs and charities have long drawn on psychological forms of expertise in their efforts to communicate with, capture the attention of and shape the choices of their audiences. More recently, the focus of such forms of expertise and techniques has been notably shifting towards the inner workings of the mind, with 'neuromarketing' consultancies promising services that can generate data on the embodied and psychophysiological responses of consumers to specific brands and products (see Schneider and Woolgar, 2012 for a critical review). Advertisements themselves have, for instance, been designed to respond to viewers' attention using eye tracking techniques – a common method in psychological research. Global behaviour change consultancies and social enterprises have emerged as important players within this industry, whilst many existing global consultancies have incorporated behavioural economics into their portfolio of services and knowledge bases (e.g. Deloitte, Gallup, KPMG, McKinsey, PricewaterhouseCoopers and RAND Corporation). There is often a close relationship between such organisations, university research centres and government-commissioned work. Some examples include Behaviour Change, Collaborative Change, National Social Marketing Centre, Nudgeathon (Warwick Business School), UCL Centre for Behaviour Change (all in the UK); iNudgeyou (Denmark); GreeNudge (Norway); Irrational Labs, The Greatest Good group (both in the USA); BEWorks (Canada); and The Behavioural Architects, Ogilvy Change (global).

Aside from advertisers, the everyday citizen-shaping domains of schooling, work and urban/building design have been shaped also by psychological and neuroscientific research (Pykett, 2015). This is further indication of the expansive and diffuse nature of psychological governance beyond the direct confines of 'the state'. Schools have long been replete with educational practices based on developmental psychology, and more recently this has been complemented by programmes based on positive psychology (Seligman, 2011), theories of 'growth mindsets' (Dweck, 2006) and character strengths (Arthur, 2005), mindfulness and 'neuroeducation' (Huppert and Johnson, 2010; Howard-Jones, 2010; see Sanchez-Allred and Choudhury, this volume). In workplaces, human resource management and employee training processes have been heavily shaped by organisational psychology, positive psychology again, psychometric testing and 'organisational neuroscience' (Becker *et al.*, 2011). And in the fields of urban design and architecture, concerted efforts have been made to establish 'neuroarchitecture' as a distinct field of research and practice to build on the foundations of work from environmental psychology and studies of spatial cognition to better design buildings and cities which respond to and potentially help to shape the psychological dimensions of human nature (Eberhard, 2009). These developments point both to the institutionalisation and formalisation of particularly psychological and biophysical accounts of human behaviour and to the way in which such knowledge is shaping governing practices within institutions (schools, workplaces, planning/architecture) beyond those of the state.

The challenges of psychological governance

It could be argued that these developments do not indicate anything particularly new about psychological governance or shaping citizenly conduct. Indeed the social and political uses (in Europe and North America) of psychological knowledge since the late nineteenth century have been well documented (Hearnshaw, 1964; Rose, 1985; Bunn *et al.*, 2001). Areas of psychiatry, education, social and family policy, criminality, military recruitment and training as well as post-conflict therapy, industrial management, Victorian self-improvement and population-based eugenic thinking have been notable in their focus on the setting of psychological norms and the training, correction and governing of minds. Some of these phenomena are directly explored by the contributors to this book. Many scholars have outlined the close interconnection between the development of academic psychology, prevailing perceptions of societal problems and the status of clinical psychology as an applied profession. Meanwhile the historiography of psychology has itself come under much criticism for its naïve search for psychology's founding fathers and its own role in sustaining a psychological discourse (Jones and Elcock, 2001; Blackman, 1994). In a radical rethinking of the history of psychology, Nikolas Rose argued that the development of psychology should not be understood as a progressive journey towards scientific truth and its subsequent application in practice, but rather that:

> [t]he conditions which made possible the formation of the modern psychological enterprise in England were established in all those fields where

psychological expertise could be deployed in relation to problems of the abnormal functioning of individuals.

(1985: 3)

It turns out that it has long been 'real' social problems which have shaped the development of psychological science rather than a clear sense of scientific development as somehow separate from the ordinary concerns of society. The promise of an evidence-based policy driven by novel scientific insight is therefore rendered problematic.

This historical account throws into question the particular idea that 'behavioural insights' are simply applied to social and governmental problems as if they were straightforward manifestations of scientific evidence. Significant and historically contingent effort has been put into the assembling of the behavioural and decision-making sciences as sciences, through the establishment of accepted methodologies, international journals, research centres and funding. It is therefore essential that we consider the ways in which activities of psychological governance are studied in their broader social, cultural, political and economic contexts; for instance, by tracing the intimate trajectories shared by neoliberal economic theories and developments in the brain and behavioural sciences (Jones *et al.*, 2013; Pykett, 2013; Davies, 2015). Contemporary manifestations of psychological governance are connected with a particularly economised vision of psychology, drawing most heavily from the discipline of behavioural economics. Like any such account, this offers only a partial view of the complexity of human behaviour, emotion, perception, cognition, intention and action. It thus arguably provides little by way of cultural, social and political-economic explanation for how and why people behave in certain ways. Indeed, as has been argued elsewhere (Tarde, 2007 [1902] in Greco and Stenner, 2013: 60), integrationist disciplines such as political economy and economic psychology (and by extension, behavioural economics and happiness economics) tend towards a split notion of disciplinary perspectives – where one discipline is said to compensate for the partiality of another. In Tarde's critical account of economic psychology, then, the work involved in constituting economics as a science serves to reformulate the 'subjectivity of desires', and the semiotics of money (as a cultural product) as abstract, measurable, objective and quantifiable. This observation is developed by Davies (this volume), who considers the inherent suspicion of discourse and language posed by contemporary happiness economics in particular.

Elsewhere (Jones *et al.*, 2013), we have outlined what we term 'the rise of the psychological state' in light of evidence of the influence of the idea of 'nudge' and behavioural science literatures on UK policy strategy and policy-making in several sectors including personal finance, environment and health. By approaching the psychological state as an anthropological phenomenon, we traced the specific enthusiasm for behavioural science explanations in policy strategy documents, white papers, think tank publications, in political discourse, by specific civil service personnel, political figures, research centres and highly publicised academics, and actually existing state practices around these three sectors. We described the appeal of 'nudge'-inspired thinking and behavioural economic thought in particular to both the modernisation of New Labour's proclaimed 'missionary style' of government, in the ascendance during the late 1990s, and the 2010 Conservative-led coalition government's emphasis on reducing bureaucracy and state expenditure. Since 2013, further evidence of the influence of the behavioural sciences on global national and supranational government strategy and policy abounds, from both critical commentators and early advocates, as already described above (see, for example, the European Commission's latest review of behavioural insights applied to policies across European nation states [Sousa Lourenço *et al.*, 2016]).

In adopting the moniker of the 'psychological state', however, there is a risk of suggesting a rather caricatured critique of government intervention as if there is something untoward about using the latest scientific evidence as a basis for designing more cost-effective policies that take into account the real complexities of the human mind and behaviour. Some critics have indeed dismissed psychological governance as being either too trivial to matter or, conversely, a radical threat to democracy achieved through malign manipulation and psychological control. However, our position has been to narrate the emergence of the psychological state as a form of political reasoning – as a set of justifications and rationales given for taking particular courses of governmental action, targeting specific individuals and social groups, and for promoting certain kinds of intervention or indeed non-intervention in specific contexts. All of these forms of reasoning are highly contentious, and it is therefore important to attempt to understand the stated rationales behind new behavioural policy-making and delivery mechanisms, the foundational knowledge bases from which they derive a certain political legitimacy, and the potential unintended consequences of their application. Far from indicating a conspiracy to govern through expert knowledge on the mind, brain, behaviour (and its relation to economic decision-making), psychological forms of governance instead raise a set of fairly normal and normative questions around how the techniques, knowledges and sources of government action are being reshaped, how the citizen-subject is reimagined through such techniques, and how deliberations around whose and what values are to be promoted as public goods, social norms and collective decision-making are reworked.

Thus a useful starting point for this book is to consider the venture of the behavioural and decision-making sciences themselves not as universal and univocal terrain but – as with science in general – a fluid and contestable set of knowledge-making practices, which can be, and are, used for a variety of political and economic ends. In the following subsections, we outline how this argument informs the structure of the book and describe the contributions of the chapters to three principal aspects of psychological governance. First we provide some commentary on the making of the discipline of behavioural economics, which underpins notions of psychological governance – exploring key debates and discontinuities that have implications for regulatory politics and state action. Second, we consider how such knowledges about the brain, mind and behaviour are connected with the particular forms of psychological governance described by the book's contributors. Third, we outline how the chapters differently approach the exercise of psychological power, its methods, tools and techniques – exploring the implications of psychological governance for social practice and citizen identity. In conclusion, we return to the question of the political significance of psychological governance with a final chapter by Peter John, who proposes an important challenge to the critical perspectives offered throughout the book.

Making behavioural economics

At a basic level, the emergence of psychological governance in public policy is closely related to the development of an 'epistemic community' of behavioural scientists whose research, insights and expertise have become highly valued amongst policy strategists and policy-makers, particularly in the UK and the USA but increasingly elsewhere. Epistemic community is a term used to describe 'a network of professionals with recognized expertise and competence in a particular domain and an authoritative claim to policy-relevant knowledge within that domain' (Haas, 1992: 3). Within an epistemic community, members have a 'shared belief or faith in the verity and the applicability of particular forms of knowledge or specific truths' (Haas, 1992: 3) and are actively involved in producing and disseminating that knowledge to address policy problems (Meyer and Molyneux-Hodgson, 2010: no pagination). The application and usability of knowledge are thus central to the belief and value systems of epistemic communities, as are their interactions across transnational networks of knowledge production and policy transfer (Haas, 1992: 4). The ways in which disciplines such as behavioural economics and, to a lesser degree, psychology and neuroscience have been used in contemporary statecraft is suggestive of a more or less coherent epistemic community. This definition is useful in unpacking the relationships between the epistemic communities that have been established around such specific disciplinary perspectives and emerging orthodoxies of what counts as evidence in psychological forms of governance.

The disciplinary evolution of behavioural economics has been essential to the emergence of psychological governance, first in the UK and the USA and subsequently in other nation states and amongst supranational organisations such as the World Economic Forum and the European Union. Its main claim as a departure from classical economic theory is that human behaviour does not fit with the abstract models on offer from the mainstream. Rather, humans are not as economically sophisticated as had been assumed; the extent of their rationality is limited and their decisions cannot be predicted by economic models alone. Only behavioural knowledge is fit to the task, and it is this kind of empirical economics that should inform policy, business practice and decision-making.

There is a tendency amongst its most famous advocates to identify the origins of behavioural economics as recently as 1980s Chicago, through the work of Richard Thaler, George Loewenstein, Colin Camerer and Robert Shiller (Thaler, 2015; Sunstein, 2015). A longer history is recounted in other 'insider' accounts, traced for instance to a key encounter in 1969 Jerusalem between two psychologists of judgement and decision-making, Amos Tversky and Daniel Kahneman (Kahneman, 2011). The latter was to go on to win the Nobel Prize in Economics in 2002 and to play an important role in influencing many of the aforementioned behavioural economists at the University of Chicago during the 1980s. Certainly, these figures have played crucial roles in the development and popularisation of behavioural economics as an epistemic community with vast impact on international public policy debates – not least through Thaler and Sunstein's publication *Nudge* in 2008, which topped bestseller lists in both the UK and the USA. Richard Thaler, along with several other influential behavioural economists, was invited to numerous seminars at the UK government's Cabinet Office, British think thanks and the OECD; and Cass Sunstein famously went on to head the US Office of Information and Regulatory Affairs between 2009 and 2012, incorporating behavioural economic approaches into regulatory policy (see Sunstein, 2011).

However, by acknowledging the much longer history of the rise of behavioural economics (Sent, 2004; Jones *et al.*, 2013; Schüll and Zaloom, 2011), we can better appreciate some of the internal wrangling that troubles the rather more coherent story offered by key advocates of the discipline. This wrangling is not merely academic but signifies something of the crucial politics of knowledge production associated with the rise of behavioural economics, which no account of psychological governance should ignore. Esther-Mirjam Sent (2004: 740–1) helpfully distinguishes between the *new* school of behavioural economics developing in 1980s Chicago and the *old* school of behavioural economics, primarily at Carnegie Tech during the 1960s but also complemented by a more international set of research clusters at Yale, Michigan, Oxford and Stirling. As we have noted elsewhere (Jones *et al.*, 2013: 4–5), the old Carnegie school was much more affiliated with the discipline of psychology than the new school, which remained firmly rooted within economics. Indeed, key thinker on the concept of 'bounded rationality', Herbert Simon (1957), through his emphasis on our psychological limits and the constraints of the sociocultural settings in which we find ourselves, unsettled the abstract models of classical economics well before the 1980s. Simon and the Carnegie school also seemed to pose much more of a challenge than new behavioural economics to the neoliberal economic thinkers who dominated the first so-called Chicago school of economics of the 1940s – figures such as Friedrich von Hayek and Milton Friedman. Their position was that 'real' human behaviour, whilst important, was not knowable or predictable, thus making markets the best impartial arbiters of personal interest and economic transaction. By contrast, Simon emphasised that bounded rationality was an important or adaptive part of the human condition, which directly problematised the notion that apparently neutral markets are the ideal system for correcting these errors and ensuring rational economic outcomes.

At heart then, there has long been a significant amount of disagreement as to the political and governance lessons one should derive from the behavioural economic principle of irrationality. Eventually, Simon was to abandon his economic project and return to the psychology department at Carnegie (Sent, 2004: 742), and the *new* incarnation of behavioural economics would not then emerge in earnest until the 1970s under the guidance of Tversky and Kahneman. Their work set out to identify patterns in the bounded rationality of human judgment and decision-making, the systematic and apparently predictable mistakes that shape our behavioural repertoires. This is perhaps the main characteristic difference between old and new behavioural economics; for the new behavioural economists influenced by Tversky and Kahneman, heuristics and biases were errors that should be predicted and rationalised. The ideal of rational economic man thus remains in place, and it is the responsibility of governments to help people to achieve and manage their rational potential. The question of who determines what constitutes a rational course of action remains, of course, a moot point (Jones *et al.*, 2013: 10).

Recalling the divergent origins of what has come to be known as behavioural economics in this way allows us to consider why the new behavioural economics of figures such as Thaler has had such appeal amongst policy-makers favouring psychological forms of governance in the UK. In one sense, it offers a potential response to some of the most apparently intransigent of global policy problems (e.g. personal indebtedness, public health issues, environmental crisis) said to require a step change in human behaviour and personal responsibility. And yet in another sense, psychological governance poses no threat to the established neoliberal economic order since it supports marketised forms of individual choice and light-touch regulatory activity. It is precisely the non-interventionist tone of psychological forms of intervention that mark out the 'behaviour change agenda' as a fruitful and somehow politically neutral form of governance.

For science and technology studies scholars and anthropologists also interested in the operation of behavioural economics as an epistemic community,

> [s]cientific debates over choice-making in the brain [...] are also debates over how to define the constraints on human reason with which regulative strategies must contend.
>
> (Schüll and Zaloom, 2011: 515)

Behavioural economics is infused with a distinctly economic account of biology, informed by a set of dualistic game theoretic accounts of internal conflict within the brain (fast/slow; emotional/rational; short-termist/future-oriented) (Schüll and Zaloom, 2011: 522; Pykett, 2013: 849). Such disciplines do not emerge in a social vacuum, of course, and the economistic bias of their biological and behavioural assertions also indicates the value ascribed to economic theory (historically above psychology) in public policy design. Economics is, after all, an *applied* discipline aimed at modelling and predicting choice rather than representing it in a *purely* scientific way. The field of neuroeconomics emerging since 2001 (using neuroscientific techniques and biological explanations to investigate and theorise economic decision-making) also has much in common with a behavioural economic concern for empirical/experimental methods and biological theory and a 'split agency' account of the human self. And yet neuroscientists themselves more often contend that the brain is a holistic system that cannot be separated into discrete parts in the way that behavioural economists and neuroeconomists tend to do (Schüll and Zaloom, 2011: 523–4). But as Schüll and Zaloom (2011: 531) point out, it is the 'two-brain model' that succeeds in gaining policy traction precisely because it does not challenge liberal democratic presumptions of individual freedom. Rather, in the new universal characterisation of human behaviour as driven by internal conflict in the mind, governance interventions need not interfere in the values and preferences of individuals. Rather, they can focus on providing the environmental cues or 'choice architectures' required to support more reflexive, rational, future-oriented, slow forms of cognition said to provide the best overall course of action. This explains why there is such a range of experience between different types of nudge. Certain nudges are aimed at cultivating deliberation and rational responses, whilst others seek to bypass conscious awareness to achieve rational responses through irrational (heuristic) means. Notwithstanding claims of the death of rational economic man, however, this position conserves the sense of human rationality on which liberal democracies are based and, thus, fits squarely *within* existing policy-making paradigms rather than posing a radical departure.

One point to take from this discussion is that the governance implications of new psychological and behavioural insights are not clean-cut, nor is their apparent novelty always as it may first appear or be narrated by its protagonists. So too is it clearly important to distinguish between sometimes subtly different epistemic communities (e.g. 'new' and 'old' behavioural economics, neuroeconomics, neuroscience or psychology) when describing the influence of particular forms of knowledge production on public policy. As such, this book is as much about the specific influence of behavioural economics on public policy (as popularised through such texts as *Nudge* and as produced through the popular imagination) as it is about other knowledge claims stemming from neuroscience, positive psychology, happiness economics and notions of psychological resilience and character.

Knowing the brain, mind and behaviour

In the next chapter, William Davies extends the above discussion of the disciplinary trajectories of behavioural economics and the behaviour change industry by turning to the growing influence of neuroscience on the psychological governance of happiness. He describes how a 'neuro-industrial complex' (Davies, this volume) intersects with developments in affective computing to render happiness an objective fact to be targeted through new kinds of government and commercial intervention. In doing so, he not only describes an under-researched aspect of psychological governance, but also identifies a strategic alliance between neoliberal states and technology companies in projecting a political project that prioritises bodily measurement over the interpretation and discussion of emotions as culturally expressed ways of feeling. This echoes Greco and Stenner's (2013) critique of the abstractive tendencies of happiness economics and what Tarde (2007 [1902] in Greco and Stenner, 2013: 60) described as a '"contempt" for lived reality itself'. New forms of (scientific, economic, computational) expertise coalesce around the visualisation of happiness via technology, measurement and metrics – clearing the ground for new forms of psychological management and normalisation based on a 'political physics' that bypasses philosophical, moral and political deliberation about how we should live. Davies' concept of 'silent citizenship' captures this bypassing of linguistic, cultural, social and political referents and their replacement with naturalised scientific accounts of human rationality, decision-making, action and morality. Thus for Davies, the psychological governance of happiness proceeds by silencing the subject – rendering people's own linguistic expressions of feeling as unreliable sources of expertise and evidence.

Yet as Davies also points out in his chapter, *talking* about psychology is, in some senses, indistinguishable from psychology itself. This is a thread picked up by Sam Binkley, whose chapter similarly explores the psychological governance of happiness. He does so from the perspective of how the pop psychology and self-help industries – key contemporary ways of talking psychologically about the self – have promoted the notion of happiness as a resource, a form of psychological capital to be maximised by enterprising subjects. These are citizens who are active, non-dependent, adaptable and perpetually changing, all characteristics apparently anathema to mutual forms of sociality and the welfare state. As with Davies, Binkley's account develops an analysis of the strategic alliance between post-Fordist forms of neoliberal capitalism and popular psychological discourses that support the emotionalisation of economic and social life. The epistemic community described here in the service of psychological governance is that of the relatively new discipline of positive psychology (in evidence since around 1997), the emergence of which Binkley describes in detail. Like Davies' analysis of happiness and behavioural economics, Binkley is careful not to overemphasise the novelty of such forms of knowledge but traces much longer-standing continuities and discontinuities in the discursive constructions of happiness and enterprise. Nonetheless, it is through these relatively recent scientific performances, practices and paraphernalia of positive psychology that happiness becomes a legitimate psycho-medical, scientific and public policy object. In the establishment of this discipline, which has aimed to counter psychology's apparent fixation on pathology and disease, new justifications are provided for self-intervention, self-directed therapeutic activity and competition within the self and with others in order to optimise one's own psychological capital and resilience. The counterpoint to these kinds of activities, Binkley argues, is the individualised sense of risk and vulnerability which necessarily accompanies the entrepreneurial practices of happiness optimisation; in a world of uncertainty, the 'individual has only her own resources to draw upon' (Binkley, this volume).

In Chapter 4, Kathryn Ecclestone provides an in-depth account of this very individualised vulnerability produced by psychological forms of governance. Her account considers happiness and behaviour change as part of the govern- ance of a wider therapeutic culture informed by discourses of wellbeing, mental health, character and resilience. She shows in particular how several successive educational, family intervention and parenting initiatives and policies in the UK since 1997 have targeted character and social and emotional learning with a view to normalising 'dispositions, attitudes and behaviours such as self-esteem, engagement, confidence, resilience, emotional management and motivation' (Ecclestone, this volume). She argues too that the behaviour change agenda itself is predicated on claims about the frailty of the human subject's decision-making capacities. Emerging from this agenda are a number of organisations (charities, campaigning groups, third sector organisations and consultancies) that make up a 'therapeutic intervention market', competing increasingly with more traditional sources of psychological expertise (educa- tional and clinical psychologists, psychotherapists and trained counsellors) to provide often short-term therapeutic programmes and packages, particularly in educational and youth work settings. Furthermore, this market, including public funding and state support for it, contributes to what Ecclestone (this volume) describes as a 'perception that psycho-emotional governance is necessary for functioning in a widening range of life situations'. In other words, it universalises and normalises vulnerability and the responsibility to relate to the self as an intrinsically vulnerable, risky and, by consequence, anxious subject. Such introspection, of course, is entirely consonant with a social and political imaginary that marginalises structural explanations for poor mental health, welfare and wellbeing. Instead citizens are expected to psychologically adapt to, accommodate and 'bounce back' from – in resilient ways – the more damaging effects of neoliberalism.

Tools, techniques and expertise in psychological governance

Running through the chapters by Davies, Binkley and Ecclestone is a critical analytical framework informed by Foucault's writings on discipline, govern- mentality and psycho-medical expertise. Davies identifies a key disconnect between attempts at disciplining psychologically through enclosed spaces and institutions via 'the micro-political physics of the body' and governing according to the 'macro-political physics of population' (Davies, this volume). This gap has been filled by the networked corporation and market techniques of surveillance and real-time performance management, buoyed by advances in digital technologies. In particular, wearable and affective computing devices that collect, store and visualise psychophysiological data pose novel ways of monitoring and maximising happiness. Key to the exercise of psychological governance in this case, therefore, is a set of technological developments that represent certain tools and techniques necessary for the practical workings of governmental power. Again this resonates with Davies' theme of silent citizenship and his contention, after Foucault, that since the late eighteenth century, we have seen a gradual and continuous shift form forms of political and moral order based in language to one seen as being derived directly from the physical, medicalised and naturalised body.

As something of a contrast, Binkley describes the tools and techniques of governmentality as operating precisely *through* language rather than circumventing it. It is thus through discursive techniques – ways of talking and writing about happiness as a cultural phenomenon – found in self-help literatures, life-coaching practices and positive psychology as a form of expert knowledge and language that the neoliberal enterprising subject is cultivated in the name of happiness. The happy, enterprising subject works on herself and her psychological capital 'on the basis of calculations of investment and return' (Binkley, this volume) and is placed in stark contradistinction to the welfare subject narrated as dependent, constrained and docile. One of the central techniques of happiness as neoliberal enterprise, as noted by Binkley and Ecclestone alike, is the reframing of psychological expertise away from therapists and into the hands of 'any organizational director (teachers, human resource managers, workplace counsellors) that inspires the self-motivated individual to undertake a set of exercises and interventions into his own mundane thought processes' (Binkley, this volume). Ecclestone in particular extends her discussion of the techniques of psycho-emotional governance with reference to the specific behaviour change agenda described at the start of this introduction. Her take on the nudge techniques pursued by the BIT in the UK is that such policy tools are predicated on highly contradictory conceptions of the human subject. On the one hand, citizens are imagined as 'subjects lacking essential psycho-emotional skills and capacities for an increasingly ruthless neo-liberal capitalist system' (Ecclestone, this volume). These subjects are in need of corrective forms of (non-)intervention that are not aimed at empowering, informing or educating but at bypassing such fragilities of decision-making. But on the other hand, nudge techniques retain a sense of optimism that the 'two-brain' citizen will somehow resolve their essential internal psychological conflicts in favour of more rational and deliberative courses of action. Furthermore, the therapeutic intervention market to which Ecclestone draws attention is based both on the theoretical splitting of the population into the ideal, functioning rational subject and his irrational, emotionally driven counterpart *and* on the normalisation of vulnerability warranting population-wide governmental strategies.

The Foucauldian analysis from which concepts of normalisation, governmentality, discourse and discipline are drawn is further shared in chapters by the Midlands Psychology Group and Gillies and Edwards, who describe how psychological knowledge has been used as a form of biopolitics in behavioural health research and as subjectification in family intervention policies, respectively. In particular, their chapters unpack the workings of psychological power and the role of measures and method in rending psychological governance practicable. In 'Psychology as Practical Biopolitics', the Midlands Psychology Group – a collective of clinical, counselling and academic psychologists founded in 2002 by David Smail (who developed a social materialist approach to clinical psychology) – describe how the psychological method links health and governance within the present neoliberal era. The chapter highlights how the measurement practices of psychology function to constitute individualised forms of subjectivity, set cultural norms relating to behaviour and the achievement of a good life and, ultimately, determine 'the right both to make live and to let die' (Midlands Psychology Group, this volume).

In focusing closely on one psychological study that examined the relationship between personality traits and a diet and exercise programme in Scotland, the chapter shows how psychometric methods and the provision of apparently neutral scientific evidence can reinforce a neoliberal vision of the responsibilised human subject within the realm of behavioural health policies ('letting die' *in extremis*). The critique here rests on a careful unpacking of the assumptions of a genre of psychometric studies that rely on self-report questionnaires and quantitative surveys as their staple methodology. The authors cast significant doubt on the validity of such methods (i.e. do they measure what they purport to?) and their reliability (i.e. do the measures consistently lead to the same results?) and, indeed, question the quantification of human experiences and character traits through the kinds of psychological scales often used in health behaviour research. Accordingly, as with the economic models of behaviour and decision-making provided by the behavioural economics described above, psychometric testing does not directly measure intrinsic psychological traits but, rather, *models* and indeed actively *constructs* such traits precisely through its own apparatus of measurement.

The chapter by Val Gillies and Rosalind Edwards likewise centres on the confluence of psychological method and knowledge with particular policy tools aimed at behavioural modification and risk prevention – in this case, within the spheres of social work and family intervention. Picking up the theme of 'character' developed in the chapters by Ecclestone and the Midlands Psychology Group, they take a more historical perspective in showing how a psychological model of character had developed in the Victorian era as a way of 'opening up mind and behaviour to public scrutiny, self-evaluation and redemption' (Gillies and Edwards, this volume). The early establishment of child protection organisations and welfare institutions in the nineteenth century, they argue, was suffused with images of sinful and degenerate parents, racialised and colonial depictions of children in need of help and 'civilisation', and '[c]hild rescue narratives' based on the premise of children's innate psychological plasticity.

Hence the exercise of psychological governance as a form of regulation of the social good through the minds and behaviour of individuals is held rather in contrast to the account set out by Davies (this volume) in which moral and cultural norms are sidelined by bodily and psychophysiological measurements purporting to get directly at a person's state of happiness, desire, preferences and wellbeing. Instead, in the case of the emergence of child psychology as a new discipline in the early twentieth century, as Gillies and Edwards recount, it is specifically moralised discourses of children's nature, potential and role in the British Empire that are mobilised in the justification of all manner of governmental interventions in family life. They go on to outline more contemporary resonances with this moralisation through an examination of political representations of the family and the construction of parenting and family life as a public concern since the emergence of neoliberal statecraft in 1980s Britain. Finally, they bring our focus back to considerations of the role of particular forms of neuroscientific disciplinary knowledge and representations of the child's brain itself as scientific justifications of policy initiatives including early intervention, the Troubled Families Unit and Family Nurse Partnerships. Once more, and in contrast to the nineteenth-century and early twentieth-century discourses on child development, psychological governance begins to circumvent political deliberation around the tenets of a good life in favour of biologized explanations for behaviour deemed irrational, unregulated emotionally or lacking in resilience.

Alberto Sanchez-Allred and Suparna Choudhury's chapter develops a number of themes raised in other contributions to the book through a discussion of their ethnographic research on mindfulness as a form of 'neuroeducation', currently proving popular in schools and youth work in North America and the UK. Rather than signifying a form of psychological governance that eviscerates moral and ethical questions on the nature of the good life (as proposed by both Davies and Binkley, this volume), they describe how the moulding of young brains through mindfulness serves to shape the brain itself as an ethical substance – a key target object for intervention in the resilience, regulation, executive function, emotional intelligence, wellbeing, positivity and character of children and adolescents. The resonances with the accounts provided by Ecclestone and by Gillies and Edwards (this volume) are clear; new forms of pedagogical, therapeutic and neuroscientific insight are being adopted as means to adjust children and young people to better withstand the demands of modern life. As such, a specific decoupling of young people's material realities from their subjectivities is in operation. Rather curiously, one could say, these tendencies actually function through the reassociation of emotional resilience and wellbeing with the materiality of the brain itself or at least, crucially, scientific and biological representations of the plastic brain as a 'site of relevant moral and pedagogical interventions' (Sanchez-Allred and Choudhury, this volume). Again, ways of talking about and narrating the material reality of the brain are important in imagining what can be done to the brain and for what kinds of ends. These very processes and mindful habits of thinking are not only significant in shaping forms of self-governance and emotional regulation, but are also implicated in the processes of future-oriented and behavioural subjectification arguably at the heart of contemporary forms of psychological governance.

Conclusion: is psychological governance 'out to get you' or is it a set of neutral policy tools?

In setting out the ways in which this book explores the tools, techniques, methods and forms of expertise associated with psychological governance in different spheres, as well as the kinds of knowledge of and epistemic communities offering insight on the brain, mind and behaviour forwarded by this phenomenon, we have proposed a critical research agenda couched in sociological, historical and anthropological perspectives on these phenomena. In this way, we have tried to put into context the specific ways in which behavioural science, behavioural economics, neuroeconomics, happiness economics, positive psychology and research on emotional wellbeing, health behaviours and child psychology are constituted by some prevailing political, moral and economic norms. So too we have tried to show how in their application, they have constituted and affected policy-making, governance and statecraft at both national and supranational scales. And finally, along with the contributors to the book, our aim is to consider the implications of psychological forms of governance for subjectivity, citizen identity and social practice.

Yet in forwarding a critical perspective, we remain mindful of the need to avoid a sense in which the accounts provided here have recourse to privileged insight into how psychological governance is currently operative in places such as the UK, the USA and elsewhere. It is important therefore not to suggest that such critical theories offer a 'big reveal' in terms of uncovering the ideological structures underpinning the behaviour change agenda, therapeutic interventions or psychological practice as biopolitics. It is necessary to acknowledge that both those actors involved in changing behaviour and those people who actively seek to change their behaviour do so according to their own normative values, reflections and critical rationalities – rather than blindly adopting the neoliberal imperatives so often spelled out in critical forms of analysis. So too is it crucial to avoid over-inflating the achievements of psychological governance, as if it is only a constraining and manipulative form of power with no recourse for contradiction, tension, conflict, resistance or progressive interpretation. As Cass Sunstein (2015: no pagination) himself has noticed:

> Some academic researchers are now falling victim to what we might call 'the Behavioral Sciences Team Heuristic,' which measures the influence of behavioral science by asking whether the relevant nation has a Behavioral Sciences Team. That's not the worst heuristic in the world, but it's pretty bad, and it often misfires. Any such team may or may not be influential

(it could even turn out to be marginal), and a lot can be, and has been, done without one.

For these and other reasons, the final chapter by Peter John offers crucial insight into the workings and remit of the UK's BIT itself from his own perspective as one of their academic advisors. He demonstrates how new policy paradigms of experimentation, testing and evaluation have provided the behaviour change industry with an evidential standard which is much desired globally. This is reflected not least in the global consultancy services now offered by the BIT. In his chapter, John questions the very notion of psychological governance and instead sets out how behavioural public policy relies on some quite standard practices, routines and values associated with the civil service. Tracing longer-running adoptions of social science in policy-making in the UK, he finds nudges and other applications of behavioural insights to be transparent, publically debated and concertedly evidence based. He suggests that it is still economics, and not psychology, that has the most bearing on public policy making.

In some ways, this is a position not dissimilar to that adopted by many of the other contributors to this book who set out the historical confluences of psychological forms of knowledge with economic theory and method and political economy. In his chapter, John focuses on empiricism rather than the political philosophy of behavioural economics as an intellectual venture, arguing that it is this empiricism which has most informed the behavioural agenda in UK public policy and, in particular, its adoption of randomised controlled trials as a mechanism for researching, evaluating and developing new policy levers. From this perspective, then, there is nothing much transformative, let alone notorious or controversial, about the BIT as a manifestation of contemporary psychological governance; rather, it offers a successful exemplar of an efficient, tested and feasible addition to a more narrow and traditional set of policy designs based around simple notions of disincentives and incentives. As John (this volume) argues, '[n]udge has been successful because it has worked within the existing agenda of state policies and according to the standard operating procedures of the bureaucracy'.

John's balanced and illuminating account provides some level-headed and pragmatic conclusions to round off the book, serving as a useful reminder of the aforementioned 'Behavioral Sciences Team Heuristic' cautioned against by Sunstein (2015). And as contributors to this book portray, it is too simple to characterise psychological governance as either instrumental or ideological. Rather, through the fine detail of their analyses, they are able to shed new light on the precise epistemic, methodological, practical and political work that has gone into assembling the phenomena we are calling psychological governance. Yet it is only in taking seriously the contention that the tools and techniques of psychological governance are somehow politically and culturally significant that we can begin to respond to some of the challenges posed at the beginning of this introduction. For even instrumental and pragmatic solutions to traditional public policy problems carry with them particular assumptions, modes of working, rationalities, partial explanations, uncertainties and unintended consequences, which require ongoing scrutiny – through academic analysis, media commentary and public and personal deliberation alike. Psychological governance is arguably neither trivial nor a radical threat to democracy but, like all forms of intervention (and non-intervention), requires justification, explanation and careful judgment. In its multiple manifestations explored throughout this book – its incantations to know oneself, maximise happiness, self-optimise or emotionally self-regulate; its behavioural modifications; and its normalisation of particular forms of subjectivity, identity and social practice – the challenges of psychological governance and the place of the citizen in her decision-making environment remain important points of enquiry.

References

Arthur, J. (2005) 'The re-emergence of character education in British education policy'. *British Journal of Educational Studies*, 53(3): 239–254.

Bason, C. (Ed.) (2014) *Design for Policy*. Gower, Farnham.

Becker, W. J., Cropanzano, R. and Sanfey, A. G. (2011) 'Organizational neuroscience: taking organizational theory inside the neural black box'. *Journal of Management*, 37(4): 933–961.

Behavioural Insights Team (2015) The Behavioural Insights Team Update Report 2013–2015. Behavioural Insights Team, London.

Blackman, L. M. (1994) 'What is doing history? The use of history to understand the constitution of contemporary psychological objects'. *Theory Psychology*, 4(4): 485–504.

Bovaird, T. and Löffler, E. (Eds) (2003) *Public Management and Governance*. Routledge, Abingdon.

Bunn, G. C., Lovie, A. D. and Richards, G. D. (Eds) (2001) *Psychology in Britain: Historical Essays and Personal Reflections*. BPS Books, Leicester.

Davies, W. (2015) *The Happiness Industry. How the Government and Big Business Sold Us Well-Being*. Verso, London.

Dweck, C. S. (2006) *Mindset: How You Can Fulfil Your Potential*. Random House, New York.

Eberhard, J. P. (2009) *Brain Landscape. The Coexistence of Neuroscience and Architecture*. Oxford University Press, Oxford.

Greco, M. and Stenner, P. (2013) 'Happiness and the art of life: diagnosing the psychopolitics of well-being'. *Health, Culture and Society*, 5(1): 53–70.

Haas, P. M. (1992) 'Introduction: epistemic communities and international policy coordination'. *International Organization*, 46(1): 1–35.

Halpern, D. (2015) *Inside the Nudge Unit: How Small Changes Can Make a Big Difference*. W. H. Allen, London.

Hearnshaw, L. S. (1964) *A Short History of British Psychology, 1840–1940*. Metheun, London.

Howard-Jones, P. (2010) *Introducing Neuroeducational Research: Neuroscience, Education and the Brain from Contexts to Practice*. Routledge, Abingdon.

Huppert, F. A. and Johnson, D. M. (2010) 'A controlled trial of mindfulness training in schools: the importance of practice for an impact on well-being'. *The Journal of Positive Psychology*, 5(4): 264–274.

Jones, D. and Elcock, J. (Eds) (2001) *History and Theories of Psychology: A Critical Perspective*. Hodder Education, London.

Jones, R., Pykett, J. and Whitehead, M. (2013) *Changing Behaviours: On the Rise of the Psychological State*. Edward Elgar, Cheltenham.

Kahneman, D. (2011) *Thinking, Fast and Slow*. Allen Lane, London. Le Grand, J. (2003) *Motivation, Agency, and Public Policy: Of Knights and Knaves, Pawns and Queens*. Oxford University Press, Oxford.

Mahony, N. (2010) 'Mediating the publics of public participation experiments', in N. Mahony, J. Newman and C. Barnett (Eds) *Rethinking the Public: Innovations in Research, Theory and Politics*. Policy Press, Bristol, pp. 15–28.

Meyer, M. and Molyneux-Hodgson, S. (2010) 'Introduction: the dynamics of epistemic communities' [online]. *Sociological Research Online*, 15(2). Available at: www.socre sonline.org.uk/15/2/14.html

Newman, J. (2001) *Modernizing Governance: New Labour, Policy and Society*. Sage, London.

Pykett, J. (2013) 'Neurocapitalism and the new neuros: using neuroeconomics, behavioural economics and picoeconomics for public policy'. *Journal of Economic Geography*, 13(5): 845–869.

Pykett, J. (2015) *Brain Culture: Shaping Policy through Neuroscience*. Policy Press, Bristol. Rhodes, R. A. W. (1997) *Understanding Governance: Policy Networks, Governance, Reflexivity and Accountability*. Open University Press, Buckingham.

Rose, N. (1985) *The Psychological Complex: Psychology, Politics and Society in England, 1869–1939*. Routledge and Kegan-Paul, London.

Sanderson, I. (2002) 'Evaluation, policy learning and evidence-based policy making'. *Public Administration*, 80(1): 1–22.

Schneider, T. and Woolgar, S. (2012) 'Technologies of ironic revelation: enacting consumers in neuromarkets'. *Consumption, Markets and Culture*, 15(2): 169–189.

Schüll, N. D. and Zaloom, C. (2011) 'The shortsighted brain: neuroeconomics and the governance of choice in time'. *Social Studies of Science*, 41(4), 515–538.

Seligman, M. (2011) *Flourish: A New Understanding of Happiness and Well-being – and How to Achieve Them*. Nicholas Brealey Publishing, London.

Sent, E.-M. (2004) 'Behavioral economics: how psychology made its (limited) way back into economics'. *History of Political Economy*, 36(4): 735–760.

Simon, H. (1957) *Models of Man: Social and Rational*. John Wiley and Sons, London.

Sousa Lourenço, J., Ciriolo, E., Almeida, S. R. and Troussard, X. (2016) *Behavioural Insights Applied to Policy: European Report 2016*. EUR 27726 EN. European Union. DOI:10.2760/903938.

Sunstein, C. R. (2011) 'Empirically informed regulation'. *University of Chicago Law Review*, 78(4): 1349–1429.

Sunstein, C. R. (2015) 'The mischievous science of Richard Thaler. Review of MIS-BEHAVING: The Making of Behavioral Economics, by Richard H. Thaler' [online]. Available at: http://newramblerreview.com/book-reviews/economics/the-mischievous-science-of-richard-thaler (accessed 29 February 2016).

Thaler, R. H. (2015) *Misbehaving. The Making of Behavioural Economics*. Allen Lane, London.

Thaler, R. H. and Sunstein, C. R. (2008) *Nudge: Improving Decisions about Health, Wealth and Happiness*. Yale University Press, London.

2 The politics of silent citizenship

Psychological government and the 'facts' of happiness

William Davies

Every year, the Oxford English Dictionary announces its 'word of the year'. In 2012 it was 'omnishambles'; in 2013 it was 'selfie'; and in 2014 it was 'vape'. In 2015, it wasn't a word at all, but a picture of a smiling face with tears running down the cheeks; the word of the year was actually an emoji. According to one linguistics expert, emoji is 'the fastest growing form of language in history based on its incredible adoption rate and speed of evolution' (Bangor University, 2015). Other forms of shorthand digitally mediated communication of emotions have arisen, such as 'TFW' (That Feeling When) and humorous images of animals accompanied by statements such as 'me when I oversleep' or 'me right now'. Thanks to these types of expression, social media platforms often serve as arenas for the parade of emotions to be graphically performed rather than discursively reported.

The rise of emoji offers one example of how the representation, communication and detection of emotions are being transformed by digital media. It is a manifestation of what has been described as the 'decline of symbolic efficiency'; that is, a perceived crisis in the capacity of verbal languages to adequately convey meaning and truth (see Žižek, 2000; Dean, 2009; Andrejevic, 2013). Coupled to this is the rise of 'empathic media' – technologies that detect emotion algorithmically via the movement of the face and eyes, bodily signals such as pulse rate or use of particular words (McStay, 2014, 2015a, 2015b). With the growth of wearable technologies, smart cities, smart homes, data mining, and so on, emotional states are becoming objects of measurement and calculation. The use of emoji is the more playful complement to the growth of this 'affective computing' complex that renders emotions visible.

Together with advances in the neurosciences, such technological and cultural developments might appear to be leading us towards a situation in which inner subjective moods can be treated as matters of *fact*. As Poovey (1998) explores, facts have a contradictory epistemological status whereby they are both independent of their context yet slot seamlessly into broader frameworks of knowledge and calculation. Thanks to the use of numbers to represent them, facts can appear 'preinterpretive or even somehow noninterpretive' (Poovey, 1998: xii). To appeal to facts is therefore to hasten the 'crisis of symbolic efficiency' inasmuch as it seeks to circumvent semiotic (and therefore cultural) forms of representation via apparently non-cultural or metacultural trans-mission systems. Objectivity surrounding emotion – one's own or that of another – is now a plausible stance to take.

This technological context provokes enthusiasm and optimism within fields of psychology, psychiatry, economics, management and market research, which aim to render happiness an object of empirical knowledge (Davies, 2015a). Together, these make up a subdomain of psychological governance that is growing in profile and power in many societies today. Since quantitative data on happiness first started being produced in the 1960s, the science of happiness has tended to be heavily reliant on subjective reports of emotions, using survey questions such as 'how did you feel yesterday?' and 'how satisfied are you with your life?'. The introduction of affect scales and questionnaires into psychiatry, for purposes of measuring depression, also relied on introspection on the part of the patient. The notion that an emotion might be detectable and measurable *without* recourse to introspection and self-reports (for instance, through fMRI, sentiment analysis algorithms, or some such) might seem to avoid methodological problems of cultural relativism. By focusing on the brain, the face or physical behaviour, scientists may be able to distinguish what someone is 'really' feeling in distinction from what they *say* they are feeling – a methodological hope that has generated particular excitement in the field of market research (e.g. Lindstrom, 2012). Happiness would become a simple matter of fact.

The spread of affective computing and empathic media into our everyday lives is therefore working in tandem with the expansion of positivist discourses surrounding human happiness. While the surging interest in happiness, positive emotion and wellbeing since the 1990s cannot be entirely explained in terms of advances in and spread of detection methods (cf. Ehrenreich, 2010; Ahmed, 2010; Binkley, 2014; Cederström and Spicer, 2015), the capacity to speak objectively and factually about happiness is dependent on the ability of certain experts to observe and monitor brains, bodies and behaviour. In that sense, this new regime of knowledge is an artefact of behaviourist and panoptical structures on which new methodologies can be built. What can begin as gimmicky demonstrations of computer wizardry can become normalised as tools of management, science and policy within a short space of time. The technical frontiers of psychological governance are moving rapidly thanks to new entanglements of the digital, behavioural and physiological.

To be sure, the project of rendering the inner self or subject empirically knowable is far from new as copious histories of the 'psych' sciences testify (Rose, 1990, 1996; Danziger, 1994, 1997; Rieber, 1980). 'Psychological govern-ment' has always involved experts with the authority and capacity to report the truth about the psychological lives of others. Meanwhile, behaviourists have been striving to circumvent subjective reports since the early twentieth century (Mills, 1998). We should be cautious of claiming some great episte-mological breakthrough in the contemporary moment, the kind of which is only likely to fuel exuberance for techniques such as neuromarketing even further. My goal in this chapter is not to suggest that current empiricist perspectives on the psyche are historically unprecedented, but to highlight some of the continuities and discontinuities between the past and present. I do, however, want to suggest that the fixation on *happiness* represents a distinct branch of psychological government, with its own privileged status within liberal politics and its own uneasy relationship to linguistic expression.

This chapter argues that the ideal of circumventing language is very long-standing, predating the birth of modern psychology in the late nineteenth century. Moreover, it should first and foremost be understood as a governmental ideal and only subsequently as a methodological ideal. For this reason, my account starts with Jeremy Bentham in the late eighteenth century, for whom an objective science of pleasure might rescue society from the threat of political and moral discourse. The question is what type of utilitarian infrastructure might support this in governmental practice. I then consider two ways in which this infrastructure has emerged: via the price system and via physiological indicators. In each case, I argue, epistemological optimism regarding happiness is based upon a combination of technological optimism and semiotic pessimism. In conclusion, I look at the limits of these forms of 'political physics', which can work so long as they narrow the field of public deliberation, but are ultimately dependent on this silencing in order to maintain their veneer of facticity.

Bentham's silent utopia

It is well known that Bentham believed that the purpose of all political authorities was to maximise the happiness of a population in a deliberate, scientific and aggregative sense. What he termed the 'principle of utility' means that all public policy interventions should be evaluated in terms of how they impact upon the happiness of society as a whole. As he stated it, 'the business of government is to promote the happiness of the society, by punishing and rewarding' (Bentham, 1988: 70). This was not, for Bentham, an ethical or political philosophy in any ordinary sense but, rather, the proposed basis of a new political science. One might even see in Bentham's thought the archetype of what modern, rational government would become.

As Foucault (2007) stresses in his lectures on governmentality, the perspective of 'government' (in contrast to that of sovereignty) is trained upon the *natural* properties and dynamics of population as opposed to their moral, political or metaphysical qualities. This is exemplified in the writings of Bentham. The starting point for Bentham's utilitarian theory is a claim about the empirics of human motivation: '*nature* has placed mankind under the governance of two sovereign masters, pain and pleasure' (1988: 1 – italics added), which emanate from the physical body into the mind. The only scientific approach to moral or political questions would be one which worked inductively from this naturalistic account of psychology and motivation. Again, this maps neatly onto Foucault's account of the roots of liberal government.

According to Foucault: 'there is at least one invariant that means that the population taken as a whole has one and only one mainspring of action. This is desire. ... Every individual acts out of desire. One can do nothing against desire' (2007: 72). The pleasure-seeking psyche is the methodological *a priori* of governmental science.

By properly understanding this psychological nature, morality and politics might be placed on a scientific footing. The moral language of 'interests', 'good', 'ought' and 'right' are, according to Bentham, metaphors whose only possible *real* referent is the happiness of the community (1988: 4). Happiness in turn is explicable only in terms of the aggregation of pleasures and pains or what Bentham termed a 'blended result' of physical stimuli (Manning, 1968). Right from the outset of this argument, Bentham warns against the illusions that language might produce to contradict him on this. 'In *words* a man may pretend to abjure [pleasure and pain's] empire: but in *reality* he will remain subject to it all the while', he argues (1988: 1 – italics added). Politics was obstructed from the discovery of scientific, rational foundations by the 'tyranny of sounds' produced by philosophers, moralists, religious authorities and traditionalists (Dinwiddy, 1989). The entire vocabulary of morality was a jumble of noises and signs, which had nothing physical to latch on to. Bentham's vision of utilitarianism promised to cleanse politics of nonsense.

Realism and semiosis are therefore pitted against one another in Bentham's analysis. It was crucial to his empiricist enterprise that conceptual and semiotic distinctions did not get in the way of identifying a single gauge of value, namely utility. Utility he defined as 'that property in any object, whereby it tends to produce benefit, advantage, pleasure, good, or happiness, (all this in the present case comes to the same thing)', with the throwaway contents of the parentheses highlighting Bentham's lack of concern with the verbal indicators of this 'property' (1988: 2). Underlying these seemingly diverse forms of experience lies the physical substrate of pleasure and pain, which generates psychological experiences that are to be communicated verbally. Maximising happiness in any scientific or evidence-based sense would necessarily involve a monistic focus upon a single index of human pleasure and pain. The alternative, from Bentham's perspective, was that politics would become once more immersed in debate and confusion surrounding the meaning of 'happiness' or 'benefit' or 'justice'.

Foucault characterises liberal government as a 'physics of power' (2007: 49) and describes discipline as a 'political technology of the body' (1991: 26). The rise of government and discipline in the late eighteenth century went hand in hand with the elimination of political metaphysics; that is, moral or religious ontologies of human virtue and freedom. For Bentham, happiness survives as the sole end of physical politics precisely because it is ultimately an effect of physiology; that is, the natural, somatic propensity to pursue pleasure and avoid pain. Utilitarianism would therefore convert morality into a variety of physics.

Various genealogies of liberal freedom illuminate the ways in which individuals come to act as particular types of moral subject. Nietzsche (2013) asked how it is that individuals are able to make promises to each other. Rose (1996) has explored how individuals come to express 'attitudes', explore their own selves in psychoanalytic dialogue or speak in 'groups'. Foucault's (2008) account of neoliberal subjectivity encompassed the ethics of entrepreneurship; his history of sexuality starts from the question of how it became possible to *speak the truth* about one's self or desires (Foucault, 1998). Liberal and psychological government needn't necessarily involve a circumvention of language in order to succeed. Certain forms of moral and epistemological speech need to be actively produced. Contemporary behaviour change policies, for example, often act via some imbued moral capacity to make spoken or written pledges for the future (Dolan *et al.*, 2010). Such cases do not operate purely on the terrain of the physical body.

The quest for political physics, of which Bentham's work is the paradigmatic example, has more specific implications for how psychological government proceeds. Crucially, the ideal of a population (or person) governed in a physical fashion involves, if not a silencing of the subject, at least a certain irrelevance of verbal self-expression. Language becomes at best a positive, albeit unreliable, indicator of some empirical reality. In the case of medicine, Foucault (1989) argues, the question 'what is the matter with you?' became supplanted in the late eighteenth century by 'where does it hurt?'. The patient speaks, but simply to refer back to the body as a matter of scientific reporting. The sufferer of pain is in any case viewed with scepticism on the basis that they are unlikely to be reporting accurately (Bourke, 2014). In psychology, this same problem would be taken up by behaviourists in the early twentieth century with the recognition that subjective reports were a regrettable if unavoidable methodological tool.

The scientific interest in happiness or utility therefore sits at the vortex of various ideals of positivism, liberalism and government. First, happiness (or desire) is taken to be the only natural basis of all ethical and political principles because it is the one psychological entity that is ultimately anchored in the body. Second, a naturalistic ethics is argued by Bentham to be a preferable alternative to philosophical or deliberative moral enquiry because it somehow avoids the perilous ambiguities of rhetoric or political discourse. Third, the authority of language is reduced to one of unreliable reporter of some internal, noncultural, physical reality. In short, the ambition to measure and govern happiness is inextricably bound up in an ambition to avoid dialogical politics. The decline of symbolic efficiency is triggered early on in the history of liberal government. From a Benthamite perspective, a rational politics would be one in which the expert gaze could observe utility as a visible, objective phenomenon in much the same way as it might observe weight or blood pressure.

Implicit within this is the notion that numbers do not suffer the same failures of representation as words. Whereas words have the tendency to mislead, to create confusion between the 'real' and the 'fictitious', numbers must have some privileged epistemological status, including in the representation and publication of inner emotional states. Benthamite utilitarianism converts all ethical questions into matters of quantitative fact: what will the effect of this action be on the aggregate happiness level? Numerical measurement is assumed to be free of the misunderstandings that plague political and moral discourse for exactly the reasons Poovey identifies in her analysis of the modern 'fact'. If happiness could be enumerated, it would be possible to speak of 'right' and 'wrong' in an objective, apolitical, trans-cultural fashion. This in turn implicitly assumes that a measuring device might exist that lacks politics or culture.

Utilitarian infrastructures

The most notorious political technology associated with Bentham's thought is the panopticon prison design, which in Foucault's (1991) genealogy, is an archetype of disciplinary power. While discipline is trained upon individual bodies, liberal government acts upon 'macro' human objects conceived in terms of population. Statistics are therefore the indispensable technology of governmental power, bringing to light the natural tendencies (health, work, birth, death, etc.) at work within populations (Foucault, 2007). Statistics make 'political physics' possible through converting otherwise moral questions of 'freedom' and 'justice' into empirical questions of phenomena such as 'public health', 'birth rate', 'economic growth', 'unemployment', and so on (Desrosières, 1998).

However, between Bentham's theory of utilitarianism – as a form of psychological government trained upon inner motives – and statistical techniques of liberal government – aimed at knowing and optimising population – there is a gap. While Bentham paid philosophical and methodological attention to the question of how to measure happiness, he offered few clues as to how the maximisation of happiness might ever become a scientific policy project in the way that, for example, the maximisation of wealth might be rendered technical by classical political economy and economic policy. This leaves the principle of utilitarianism at the status of a heuristic device for policy-makers to keep in mind, but rarely to employ as a technique or measure of power (Quinn, 2013). Speculatively speaking, how *might* a science of population be combined with a psychological science of utility?

Bentham did offer methodological clues for how the detection of inner psychic states might be achieved. Aside from more straightforward means of quantification (such as the duration or probability of a pleasure), he identified two possible ways of representing the intensity of pleasure: pulse rate and money (McReynolds, 1968). The birth of modern experimental psychology in the late nineteenth century, organised around laboratories, introduced additional means of trying to gauge psychological states via observation, such as those which focused on the eyes (Danziger, 1994; Crary, 2001). But for the purposes of a scientific, utilitarian government, attuned to the fluctuations of pleasure and pain, the challenge is not to build further enclosed spaces (such as panopticons) but to build large-scale public infrastructures of happiness detection. Arguably, it is just this sort of infrastructure that contemporary affective computing technologies promise to deliver, as manifest in contemporary experiments in the gauging of public emotion via the face. Yet the blueprint for utilitarian infrastructures can be traced back to Bentham's identification of pulse rate and money as the two likeliest indicators of pleasure and pain. Let's consider each of these separately.

Happiness via price

Bentham intuited that money might serve as a workable proxy for happiness. As he explained:

> Speaking of the quantity (the respective quantities) of various pains and pleasure and agreeing in the same propositions concerning them ... we must make use of some common measure. The only common measure the nature of things afford is money. How much money would you give to purchase such a pleasure? 5 Pounds and no more. The two pleasures are (much as to you be reputed) equal.
>
> (Quoted in McReynolds, 1968: 353)

The insight that psychic pleasures and pains might have some natural relationship to money would later serve as the basis for English marginalist economics and welfare economics. During the 1870s and 1880s, William Stanley Jevons and Francis Edgeworth built their neoclassical economic theories upon the assumption that consumers would exchange money for proportionate quantities of utility in a rational, calculated fashion (Schabas, 1990; Sigot, 2002; Maas, 2005). Edgeworth went as far as arguing that economic analysis might one day employ a 'hedonimeter', which could measure fluctuations in happiness across the population, and use this to understand changes in market prices (Colander, 2007). Contemporary happiness economics has picked up Edgeworth's technical project with at least two happiness measures bearing the name 'hedonimeter'.[1]

From Bentham's perspective, the significance and benefit of the market price system was that one could safely assume that it was optimising utility without requiring any further psychological science. If utility is believed to be in a faintly stable relationship with money, then a science of price will act as an effective proxy for a science of happiness. By the 1890s, this was also the view adopted by neoclassical economists such as Marshall and Pareto who argued that economists needn't get involved in speculative or psychological questions about how much pleasure goods might provide but, rather, could focus on choices between rival goods (Hands, 2009). And by the 1930s, this had turned into the behaviourist tautology of 'revealed preference' theory – the idea that if consumers exercise a choice, they must *de jure* prefer this option over others. The psyche fell out of neoclassical economics altogether.

The market might therefore serve as a utilitarian public infrastructure whose prices could serve as a real-time indicator of wellbeing. It is implicit within this vision that the market does not simply have an 'economic' function in distributing material goods, satisfying needs or increasing production, but a liberal political function in mediating diverse human desires and differences. If we follow Bentham's logic, it is preferable to govern society on the basis of some form of 'revealed preference' than to do so on the basis of expressed, verbalised preference. The former is more *realistic* than the latter, which tends to fall victim to the tyranny of sounds. Consumer behaviour, revealed in the fluctuation of prices, becomes the object of psychological government. From this perspective, even the market research industry might be able to claim that it is acting in a utilitarian fashion, once all market behaviour is assumed to be utility-maximising.

The understanding of the market as a political mechanism superior to democracy would later serve as one of the central tenets of Chicago school neoliberalism, partly inspired by Friedrich Hayek. Bentham was a clear inspiration in the combining of pessimism towards deliberative politics with optimism regarding scientific alternatives; the Chicago school's debt to Bentham was sometimes quite explicit (Kitch, 1983; Engelmann, 2003). Milton Friedman (1953) argued that moral differences will only ultimately be resolved in violence – and not in discourse – making it imperative that positivist or market-based solutions be found. Gary Becker suggested that 'there is relatively little to choose between an ideal free enterprise system and an ideal political democracy; both are efficient and responsive to preferences of the "electorate"' (1958: 108). But this equivalence only makes sense on the assumption that both market and democracy are utilitarian infrastructures – techniques for the mass auditing of desire.

The risk with this form of neoliberalism is that it ultimately produces another form of dogmatism in which there is no social or political question that cannot be resolved via economics and markets. Once price is granted such a privileged status in the representation of desire – a status first bestowed by Bentham – there is no limit to what economics might process or what markets might be able to do. This 'economic imperialism' immediately invites the question: what if some forms of value are not captured by price? And what if individuals do not express their desires or goals in market behaviour? Questions of the psyche return to economics once more, as has occurred in the rise of happiness and behavioural economics from the 1970s onwards.

Happiness via diagnosis

The interest in pulse rate has a lengthy history, dating back to ancient times; although it wasn't until the modern era that pulse rate began to be measured through counting the number of beats in a minute (Ghasemzadeh and Zafari, 2011). Bentham recognised that pulse rate may offer an indication of inner psychological states, but he was not optimistic about the capacity of this measurement to provide a scientific basis for utilitarianism. The rise of wearable technology in the early twenty-first century has seen pulse rate become a more significant indicator of wellbeing, being linked to levels of stress, physical exercise and sleep quality. But until this development, physiological signs and symptoms could only be known via 'macro' statistics on mortality rates and specific diseases.

The suggestion that pulse rate might act as a measurable proxy for pleasure may have been somewhat far-fetched in Bentham's time, but it did nevertheless point to a utilitarian agenda that acts in parallel to that of market liberalism. Experiments in 'psychophysics' conducted by Gustav Fechner and others from the 1840s onwards sought to develop a scientific psychology on the back of physiology. Expert observation and diagnosis of the body *could* become a basis on which to evaluate the psychological wellbeing of a population, not just the physical wellbeing. This type of utilitarian infrastructure potentially includes: techniques developed by behavioural psychologists trained upon the face, eyes and body; neuroscientific scanning of the brain via fMRI or EEG; or medical examinations of the body. None of these need be put in the service of monitoring or maximisation of happiness, but they may be. In the case of medical expertise, 'quality of life' became an explicit concern of healthcare decisions and policies from the 1970s onwards (Benzer, 2011).

In Foucault's historical analysis of modernity, one of the distinguishing features of such techniques of observation and measurement is that they exist within enclosed spaces; that is, within institutions of *discipline*. This is not to say that discipline works according to a utilitarian logic of acting on happiness; on the contrary, Foucault suggests that disciplinary institutions such as prisons, schools and the military have a normalising function rather than an optimising function. Bentham's hope was that such institutions of normalisation would be put in the service of happiness maximisation at the aggregate, societal level (for instance, prison being sufficiently unpleasant that the crime rate would remain happily low). Yet until the rise of methodologies such as attitudinal surveys in the 1920s (Rose, 1996), government would have considerable means of acting upon the psychology of those within disciplinary institutions but little means of monitoring the psychology of those *outside* disciplinary institutions. And while surveys facilitated new forms of psychological government, such as those trained upon 'opinion' or 'morale' (Osborne and Rose, 1999), they remain stuck with the Benthamite problem of people misreporting their inner states, motives and desires. Moreover, they operate only periodically, meaning that fluctuations of mood cannot be very closely monitored in the way that a patient might be within a hospital or a subject within a psychology laboratory.

One way of framing this problem is in terms of the disjointed relationship between 'governmentality' and 'discipline' where the former involves a form of macro-political physics of population and the latter, a form of micro-political physics of the body. Whereas the market promises to both monitor and aggregate internal psychological states, techniques for monitoring and optimising the body tend to be confined to enclosed spaces; they do not scale up of their own accord in the way that prices are able to do.

This is where Deleuze's (1992) analysis of 'societies of control' becomes so significant to the analysis of psychological government. Disciplinary societies involve a polarity (though not a contradiction), Deleuze argues, between the individual and the 'mass'. They are structured around a series of confined institutional spaces through which individuals would pass as members of the 'mass': school, hospital, barracks, factory, laboratory, and so on. The logic of 'control', however, works between the poles of individual and mass, producing networks and perpetual connectivity which engulf various scales of population. 'In the disciplinary societies one was always starting again (from school to the barracks, from the barracks to the factory), while in societies of control one is never finished with anything' (Deleuze, 1992: 5).

In practice, this means that techniques of behavioural or physiological observation that might once have been confined to enclosed spaces, such as doctors' surgeries or psychology laboratories, are now spread across public spaces too. Or to put that another way, properties that were once unique to the market (i.e. the capacity to process information and desires in real time and outside of any formal institutional structure) are now being acquired by techniques of surveillance and expertise (Davies, 2015b). The combination of social media with smartphones provides one contemporary manifestation of how the logic of control expands, starting within the confines of the corporation (as an ideal of real-time performance management) and then spreading beyond into day-to-day social relations. The capacity of the corporation (described by Deleuze as a 'gas') to penetrate more and more areas of social life is at the heart of how control has come to usurp discipline.

Bentham's intuition that pulse rate might play a role in the science of pleasure was actually highly prophetic. A *public* utilitarian infrastructure built around movements in and of the body may not have been easily conceivable for Bentham, who was striving to design a society of discipline, but it has become realised in the context of cybernetic projects characteristic of societies of control (Franklin, 2015). 'Digital healthcare', wearable technology, smart devices, affective computing – all now offer the possibility of measuring well-being via physiological proxies across the population at large in real time (Lupton, 2012, 2013a, 2013b). Ubiquitous digitisation and the 'internet of things' makes possible new forms of algorithmic governance in which complex systems can be steered on the basis of continuous, real-time feedback to identify new patterns and correlations (Dormehl, 2014). Beyond just pulse rate, other quantitative physiological indicators are now traced, such as respiration rate and steps taken. Workplaces are one centre for these new forms of psychological and physiological government; here, data is generated that shows fluctuations in happiness and wellbeing from moment to moment, just as the price system has been alleged to. In that sense, new happiness-monitoring infrastructures are challenging the neoliberal insistence that only the market can rescue us from the tyranny of sounds and political violence.

As with the pleasure-seeking consumer in the marketplace, the digitally enhanced cyborg body has no need to speak. The fantasy of a utilitarian society is realised in this instance through constant monitoring of physical movement and behaviour. The body (including the brain, face, pulse rate, and so on) comes to 'speak' in ways that are less susceptible to semiotic and cultural deception than verbal expression. 'Smart' environments, capable of reading signals and quantities transmitted by physical bodies, provide one route to the utilitarian liberal ideal in which psychological science (albeit of a resolutely behaviourist variety) might be married to the government of population.

The limits of 'political physics'

The science and government of happiness rests on a profound ambiguity that produces a constant tension in how it approaches its object. On the one hand, happiness is posited as the 'ultimate' goal of human decision-making. In Bentham's terms, this is because pleasure and pain are the two 'sovereign masters' that nature has established over us. Such claims are treated as beyond question by Benthamites. The happiness economist and policy advisor Richard Layard has written that 'if we are asked why happiness matters, we can give no further external reason. It just obviously does matter' (2005: 113). In order to succeed politically and ethically in overcoming conflict and debate, happiness has to be treated as a quasi-transcendent telos that nobody can reject. It is worth noticing that, as far as Bentham is concerned, anyone who claimed not to be driven by pleasure and avoidance of pain (the components of happiness) would not be worth listening to anyway.

But in order to serve as an object of natural and behavioural science, happiness needs to be substantially less than this 'ultimate' goal. It needs to be signified in relatively prosaic acts of consumer preference or bodily rhythms. These types of indicator may not be treated as identical to the underlying substance that they are believed to represent. But then, what is the status of that underlying substance if it is not simply another prosaic form of physical matter? The rise of the neurosciences has made this issue more acute, displacing 'mind' with 'brain' but immediately confronting the question of why the brain matters especially if the mind has been displaced. As Rose argues, 'the deep psychological space that opened in the twentieth century has flattened out' (Rose, 2007: 192). Without that 'deep psychological space' lying beneath indicators such as the price system or physiology, it is not clear why either would be deemed politically or normatively important.

The changing definition and government of depression since the 1960s is one of the key areas where this ambivalence has been played out (Ehrenberg, 2010). Attempts to place psychiatric diagnosis on a consistent, naturalistic, scientific footing have transformed expert and governmental perspectives on mental illness generally, with the rise of the 'neo-Kraepelinian' school of psychiatry focusing on quantifiable behavioural cues (Decker, 2013). This has been crucial for bringing mental illness within managerial and governmental frameworks of risk and cost-benefit analysis. But especially in the case of depression, there are often philosophical quandaries surrounding what counts as a 'disorder' and what are the 'symptoms' of that disorder. Unhappiness is both the underlying condition and the observable phenomenon at the same time. Symptoms (such as insomnia or loss of sex drive) can swiftly turn into disorders in their own right once there is no clear distinction between diagnostic signifier and signified. This problem arises out of the philosophical ambiguity of treating something like unhappiness as a natural fact.

The dilemma of Benthamism, and by implication of utilitarian government, is that happiness needs to be both an ethical substance and a matter of natural fact at the same time. As Foucault argues with respect to disciplinary punishment, physical contact with the body only occurs so as 'to reach something other than the body itself' (Foucault, 1991: 11). Political physics and biopolitics attempt to operate within the plane of positive empiricism but end up attributing non-empirical, political qualities to those aspects of behaviour or of the body that are made visible by utilitarian infrastructures. This is also the charge that is made by Bennett and Hacker against neuroscience, which they claim, 'ascribes to the brain much the same range of properties that Cartesians ascribe to the mind' (2003: 111).

This critique can be made with respect to both of the utilitarian infrastructures examined in the previous section. Once the price system becomes viewed as the indicator of moral value, and not only of material preference, it becomes imbued with a fundamental political status as the constitutional basis of liberalism. This is precisely the problem that neoliberalism encounters where it comes to inflate technocratic and positivist agendas with metaphysical forms of authority (Davies, 2014). Similarly, once bodily symptoms and rhythms become viewed as indicative of moral value, and not only of physiological health, they take on an ethical and existential status. This is what the critique of 'healthism' or 'somatopia' seeks to highlight (Crawford, 1980; Lupton, 1995; Chrysanthou, 2002).

Psychological empiricism does not succeed in avoiding dilemmas of Cartesian dualism. The measurement of happiness does not coherently avoid normative or metaphysical questions regarding how one ought to act; it simply imposes strictures around how one might authoritatively speak about that question. Pragmatically speaking, this translates into an attempt to close down public deliberation so as to reduce the number of people permitted to speak on questions of ethics. Questions of the 'good life' become amenable to technocracy. The science and government of happiness is impossible without the downgrading and disciplining of critical and evaluative voices that would otherwise speak via multiple (though not necessarily incommensurable) registers and rhetorics of valuation. Dismissing this pluralism as the tyranny of sounds or the road towards violence allows the happiness scientist to offer their methodological infrastructure as the safer empirical path to legitimate public decision-making.

The philosophical alternative to this form of government lies in a different understanding of psychological language altogether. From a Wittgensteinian perspective, the language of psychology is unusual in being indistinguishable from that which it seeks to describe. Expressions of one's happiness or pleasure or pain are not attempts to represent some set of facts within the mind or body and cannot be judged in terms of their empirical validity; rather, they constitute meaningful, comprehensible emotional behaviour (Harré and Secord, 1972; Bennett and Hacker, 2003; Wittgenstein, 2001). Such behaviour is meaningful inasmuch as it makes sense in the social context in which it is located. Something like 'pain' is a matter of interpretation, but for the same reason, the manner of its description is crucial to what it means and is. The terms that are used to express an emotion are not clumsy, avoidable empirical indicators of some inner fact, but *ways of being emotional.*

To return to the question of the decline of symbolic efficiency, where this chapter started out, this decline is never quite what it seems. The rise of empathic media and the visualisation of moods ostensibly circumvent language, rendering emotions such as happiness matters of fact. Yet, for all the rhetorical extravagance of the neuro-industrial complex, one never actually sees the *mood itself* or measures *happiness itself*, and the claim that one might do so is riddled with philosophical problems, chief among which is a constant flipping of Cartesian dualism from positivism to metaphysics and back again. What one sees when a 'brain lights up' or when there are variations in pulse rate or fluctuations in the stock market is neither a simple objective psychological fact nor some underlying substance. It is a medium of cultural expression to be interpreted and understood.

The rise of emoji as the fastest growing language in the world, together with the playful social media use of animal images to represent how one is feeling, offers an ironic manipulation of the excesses of psychological government and neurological positivism. On the one hand, emoji is an *example* of the decline of symbolic efficiency, as I argued at the outset. It appears like an avoidance of verbal discourse, as if a smiley face offered immediate access to an emotion of a sort that the words 'I'm so happy' does not. And yet, any brief encounter with the mores of day-to-day digital social life demonstrates that cultural symbolism is far from circumvented. To send someone a picture of a face with tears rolling down its cheeks is not to state a fact that one is currently in tears. On the contrary, it is just another playful linguistic expression, which in its more mischievous manifestations, may even represent a sarcastic critique of the positivist psychological claim that it can ever be possible to capture emotions in a naturalistic fashion.

Note

1 See www.mappiness.org.uk, which uses an iPhone app to collect happiness data, and www.hedonometer.org, which measures the happiness of Twitter using a sentiment analysis algorithm.

References

Ahmed, S. (2010) *The Promise of Happiness.* Duke University Press, Durham, NC.
Andrejevic, M. (2013) *InfoGlut: How Too Much Information is Changing the Way We Think and Know.* Routledge, New York.

Bangor University (2015) 'Emoji "fastest growing new language"' [online] *Bangor University*. Available at: www.bangor.ac.uk/news/university/emoji-fastest-growing-new-language-22835 (accessed 24 February 2016).

Becker, G. S. (1958) 'Competition and democracy'. *Journal of Law and Economics* 1: 105–109.

Bennett, M. R. and Hacker, P. M. S. (2003) *Philosophical Foundations of Neuroscience*. Wiley, London.

Bentham, J. (1988) *The Principles of Morals and Legislation*. Promethean Books, Buffalo, NY.

Benzer, M. (2011) Quality of Life and Risk Conceptions in UK Healthcare Regulation: Towards a critical analysis. Centre for Analysis of Risk and Regulation (CARR) Discussion Paper DP 68. London School of Economics and Political Science, London.

Binkley, S. (2014) *Happiness as Enterprise: An Essay on Neoliberal Life*. SUNY Press, Albany, NY.

Bourke, J. (2014) *The Story of Pain: From Prayer to Painkillers*. Oxford University Press, Oxford.

Cederström, C. and Spicer, A. (2015) *The Wellness Syndrome*. Polity, London.

Chrysanthou, M. (2002) 'Transparency and selfhood: utopia and the informed body'. *Social Science and Medicine*, 54(3): 469–479.

Colander, D. (2007) 'Retrospectives: Edgeworth's hedonimeter and the quest to measure utility'. *Journal of Economic Perspectives*, 21(2): 215–226.

Crary, J. (2001) *Suspensions of Perception: Attention, Spectacle, and Modern Culture*. MIT Press, Cambridge, MA.

Crawford, R. (1980) 'Healthism and the medicalization of everyday life'. *International Journal of Health Services*, 10(3): 365–388.

Danziger, K. (1994) *Constructing the Subject: Historical Origins of Psychological Research*. Cambridge University Press, Cambridge.

Danziger, K. (1997) *Naming the Mind: How Psychology Found Its Language*. Sage, London.

Davies, W. (2014) *The Limits of Neoliberalism: Authority, Sovereignty and the Logic of Competition*. Sage, London.

Davies, W. (2015a) *The Happiness Industry: How the Government and Big Business Sold us Well-being*. Verso, London.

Davies, W. (2015b) 'The chronic social: relations of control within and without neoliberalism'. *New Formations*, 84–85: 40–57.

Dean, J. (2009) *Democracy and Other Neoliberal Fantasies: Communicative Capitalism and Left Politics*. Duke University Press, Durham, NC.

Decker, H. S. (2013) *The Making of DSM-III: A Diagnostic Manual's Conquest of American Psychiatry*. Oxford University Press, Oxford.

Deleuze, G. (1992) 'Postscript on the societies of control'. *October*, 59: 3–7.

Desrosières, A. (1998) *The Politics of Large Numbers: A History of Statistical Reasoning*. Harvard University Press, Cambridge, MA.

Dinwiddy, J. R. (1989) *Bentham*. Oxford University Press, Oxford.

Dolan, P., Hallsworth, M., Halpern, D., King, D. and Vlaev, I. (2010) MINDSPACE: Influencing Behaviour for Public Policy. Institute for Government/Cabinet Office, London.

Dormehl, L. (2014) *The Formula: How Algorithms Solve all our Problems … and Create More*. W. H. Allen, London.

Ehrenberg, A. (2010) *The Weariness of the Self: Diagnosing the History of Depression in the Contemporary Age*. McGill-Queen's University Press, Montreal.

Ehrenreich, B. (2010) *Smile Or Die: How Positive Thinking Fooled America and the World*. Granta Books, London.

Engelmann, S. (2003) *Imagining Interest in Political Thought: Origins of Economic Rationality*. Duke University Press, Durham, NC.

Foucault, M. (1989) *The Birth of the Clinic: An Archaeology of Medical Perception*. Routledge, Oxford.

Foucault, M. (1991) *Discipline and Punish: The Birth of the Prison*. Penguin, London.

Foucault, M. (1998) *The History of Sexuality: The Will to Knowledge: Volume 1*. Penguin, London.

Foucault, M. (2007) *Security, Territory, Population: Lectures at the Collège De France, 1977–78*. Palgrave Macmillan, Basingstoke.

Foucault, M. (2008) *The Birth of Biopolitics: Lectures at the Collège De France, 1978–79*. Palgrave Macmillan, Basingstoke. Franklin, S. (2015) *Control: Digitality as Cultural Logic*. MIT Press, Cambridge, MA.

Friedman, M. (1953) *Essays in Positive Economics*. University of Chicago Press, Chicago.

Ghasemzadeh, N. and Zafari, A. M. (2011) 'A brief journey into the history of the arterial pulse'. *Cardiology Research and Practice*, 2011: 1–14. DOI:10.4061/2011/164832.

Hands, D. W. (2009) 'Economics, psychology and the history of consumer choice theory'. *Cambridge Journal of Economics*, 24(4): 633–648.

Harré, R. and Secord, P. F. (1972) *The Explanation of Social Behaviour*. Basil Blackwell, Oxford.

Kitch, E. W. (1983) 'The fire of truth: a remembrance of law and economics at Chicago, 1932–1970'. *The Journal of Law and Economics*, 26(1): 163–234.

Layard, P. R. G. (2005) *Happiness: Lessons from a New Science*. Allen Lane, London.

Lindstrom, M. (2012) *Buyology: How Everything We Believe about Why We Buy is Wrong*. Random House, London.

Lupton, D. (1995) *The Imperative of Health: Public Health and the Regulated Body*. Sage, London.

Lupton, D. (2012) 'M-health and health promotion: the digital cyborg and surveillance society'. *Social Theory and Health*, 10(3): 229–244.

Lupton, D. (2013a) Digitized Health Promotion: Personal Responsibility for Health in the Web 2.0 Era. Sydney Health and Society Group Working Paper No. 5. Sydney Health and Society Group, Sydney, Australia.

Lupton, D. (2013b) 'Quantifying the body: monitoring and measuring health in the age of mHealth technologies'. *Critical Public Health*, 23(4): 393–403.

Maas, H. (2005) *William Stanley Jevons and the Making of Modern Economics*. Cambridge University Press, Cambridge.

McReynolds, P. (1968) 'The motivational psychology of Jeremy Bentham: II. Efforts toward quantification and classification'. *Journal of the History of the Behavioral Sciences*, 4(4): 349–364.

McStay, A. (2014) *Privacy and Philosophy: New Media and Affective Protocol*. Peter Lang, Oxford.

McStay, A. (2015a) 'Now advertising billboards can read your emotions … and that's just the Start'. *The Conversation*, 4 August. Available at: http://theconversation.com/now-advertising-billboards-can-read-your-emotions-and-thats-just-the-start-45519

McStay, A. (2015b) 'Soon smartwatches will listen to your body to work out how you're feeling'. *The Conversation*, 10 March. Available at: http://theconversation.com/soon-smartwatches-will-listen-to-your-body-to-work-out-how-youre-feeling-38543

Manning, D. J. (1968) *The Mind of Jeremy Bentham*. Longmans, London.

Mills, J. (1998) *Control: A History of Behaviorism*. NYU Press, New York.

Nietzsche, F. (2013) *On the Genealogy of Morals*. Penguin, London.

Osborne, T. and Rose, N. (1999) 'Do the social sciences create phenomena? The example of public opinion research'. *British Journal of Sociology*, 50(3): 367–396.

Poovey, M. (1998) *A History of the Modern Fact: Problems of Knowledge in the Sciences of Wealth and Society*. University of Chicago Press, Chicago.

Quinn, M. (2013) *Utilitarianism and the Art School in Nineteenth-Century Britain*. Pickering & Chatto, London.

Rieber, R. W. (1980) *Wilhelm Wundt and the Making of a Scientific Psychology*. Plenum Publishing Company Limited, New York.

Rose, N. (1990) *Governing the Soul: The Shaping of the Private Self*. Routledge, London.

Rose, N. (1996) *Inventing Our Selves: Psychology, Power, and Personhood*. Cambridge University Press, Cambridge.

Rose, N. (2007) *Politics of Life Itself: Biomedicine, Power and Subjectivity in the Twenty-first Century*. Princeton University Press, Princeton, NJ.

Schabas, M. (1990) *A World Ruled by Number: William Stanley Jevons and the Rise of Mathematical Economics*. Princeton University Press, Princeton, NJ.

Sigot, N. (2002) 'Jevons's debt to Bentham: mathematical economy, morals and psychology'. *The Manchester School*, 70(2): 262–278.

Wittgenstein, L. (2001) *Philosophical Investigations: The German Text with a Revised English Translation*, third edition. Blackwell, Oxford.

Žižek, S. (2000) *The Ticklish Subject: The Absent Centre of Political Ontology*. Verso, London.

3 Happiness as resource and resilience

An emotion for neoliberal times[1]

Sam Binkley

For a long time, conventional wisdom has understood modern economy as a threat to emotional life. Whether in the realms of production or consumption, such effects as rationality, repetition, commodification and alienation have been thought to reduce emotional existence to drab and dreary states of being, in contrast to the presumably authentic and vital forms experienced in nature or in preindustrial life. This thesis, of course, has been rendered obsolete as new forms capital have evolved around our personal, affective and emotional capacities (Illouz, 2007). The diminishment of the institutional form of bricks-and-mortar capitalism that prevailed under post-Fordist, neoliberal, and late capitalism has not only allowed the recognition of emotion in the lives of workers, consumers and citizens but has discovered the value of emotions in capitalist enterprise itself, organized around the imperatives of a service industry. Emotions now have value as a component of "human capital": today we speak of emotional labor, emotional capitalism and a broader "emotionalization" of economic and social life – a process that demands that emotions be produced, cultivated and managed opportunistically in new and aggressive ways (Holmes 2010, 2015). Emotions are not only labor commodities to be sold, but also social assets to be refined and communicated – objectives for which new forms of self-help and popular psychology stand ready to lend assistance.

Moreover through emotionalization processes, emotions appear not only as economic and cultural assets to be cultivated, but as fundamental existential resources to be honed and mobilized in the face of the risks and uncertainties encountered in everyday life. Emotions have become expressive instruments of advancement and agency but also shields or defenses against threats emerging from an uncertain and changing world. Recent critical studies of resilience have uncovered the ways in which a reflexive shaping of emotional life aims to cultivate a quality of durability and elasticity within the contemporary subject through a popular self-help genre that draws on psychological research into human resilience (Reid, 2012; Evans and Reid, 2014). In what follows, a critical engagement with the recent focus on happiness explores links between happiness discourse, and specifically the popular focus on positive psychology, and the resilience imperative, as expressed in an emerging body of self-help literature. Happiness is considered through the lens of recent critical work on neoliberalism, as taken up in studies of governmentality where happiness appears as an instrument or resource for the enterprising subject. Drawing on the author's recently published work on happiness and neoliberal enterprise, the problem of happiness is considered as a technology of enterprise and opportunity (Binkley, 2014). However, while happiness might be coextensive with the risk-taking imperatives of enterprise, it also moves the subject in precisely the opposite direction. While it edifies risk, happiness also adopts a defensive posture relative to a vulnerable subject grasping for resilience.

Happiness, resilience and neoliberal governmentality

The recent appearance of happiness as a problem of psycho-medical discourse asks us to considerably broaden our understanding of medicine itself (Binkley, 2014; Zevnik, 2014; Ahmed, 2008). Happiness is understood as a natural expression of human vitality – not an affliction or a deformity experienced by an unfortunate few but a future, a potential for the full development of a vital capacity shared by anyone and everyone. Indeed, the new technology of happiness is not just something for the clinically depressed, the deranged or other persons marked for psychological marginalia (such individuals are referred back to the old disciplinary apparatus, which stands patiently on the sidelines for this purpose). Happiness is for the average, the common, the unafflicted – those who simply want more out of life. The problem of happiness, therefore, is the perfect mechanism for ensuring that everyone becomes a psychological subject. It is, in a sense, the democratization of psychological life – one need no longer be sick to be psychological, though it is a form of democratization that brings with it many concealed and coercive effects. Anyone who falls short of the full realization of their happiness potential, happiness experts argue, has betrayed his or her own most implicitly human capacities. Not just unhappiness, but the failure to be as happy as possible, has become problematic. Such is the blackmail of happiness: to choose not to be happy is to choose against one-self and against the mandate of biological life, what one is and what one might become, which is an unthinkable choice (one only possible for those afflicted not necessarily with depression but with the malaise of everyday pessimism).

Thus, the effects of happiness extend far beyond the traditional domains of the therapist and the psychiatrist, wardens of those subjects whose states of compromised mental health, contorted by disease and maldevelopment, have long held the clinical gaze of the hospital and the asylum. Happiness is the problem for people who don't have a problem. It is not an abnormality one discovers within oneself through techniques of introspection and self-assessment in the closed spaces of clinics and asylums but is, instead, a potential to be developed in the open spaces and otherwise healthy moments of everyday life. Since happiness has no other purchase on the subject than its immediate experience of its own well-being (which is, of course, immediately transparent to every psychological subject) and its potential to maximize this well-being, it speaks to life's very movement, to the forward thrust of life itself, to the subject's vitality and ultimate capacity for a richer, fuller, happier life. To seek after happiness is to empower oneself, in the sense that Barbara Cruikshank uses the term to signify a new technology of government and a mode of subjection that is at once voluntary and coercive: "The will to empower ourselves and others," writes Cruikshank, "has spread across academic disciplines, social services, neighborhood agencies, social movements, and political groups, forging new relationships of power alongside new conceptualizations of power" (1999: 72).

As such, happiness imposes a fundamental transformation of the psychological problematic itself where democratization and empowerment occur through a unique temporalization of the problem of psychological health. The uncertain status of happiness appeals to the unfolding dynamic of a vital process, grafting itself onto the time of our life trajectories and everyday conducts, fusing with the thrust of our life energies, charting a future, a hope, a potentiality and a horizon of endlessly optimizing capacities and endlessly enriching experiences. This is not the temporality of the psychotherapist or the analyst who searches the past for the buried causes of present dilemmas and whose aim is the prescription of a cure. It is a temporality that looks to an open future of ongoing possibilities, that strategizes and seeks opportunities for ever greater utility and higher emotional returns on life's investments. Today's happiness is the temporality of enterprise.

Happiness asks us to train our eyes on a horizon of possibility and to pose the problem of our lives and our identities within an engineered trajectory of measurable risk and uncertainty, a cost-benefit analysis whose unfolding is directed by our own competencies, capacities, resources and choices, leading to the uncertain realization of our potential for fulfillment. In this way, happiness reconstitutes identity and emotional well-being as a problem not of a search for origins but of the optimal exploitation of environmental resources, opportunities and enterprises, confronted in the here and now of personal life. Moreover, happiness, as life lived to the fullest, applies a maximizing logic to those vital forces that define the very dynamism of our biological existence. Happiness is what we experience when our life forces are fully activated – to deny happiness is to deny what we are as living entities. For this reason, I argue that the new discourse on happiness affects an intensification of the apparatus of the psy-disciplines – a shedding of its heavy institutional form that enables a penetration of power going beyond our bodies and behaviors to touch on our very potentialities, futures and temporalities as subjects.

Empirically, it is possible to speak of the new discourse on happiness on a number of levels, not all of which cohere around a single genre. This new discourse is largely interdisciplinary, spanning scientific, economic, policy, journalistic and popular cultural genres, all of which exert a combined influence on lay and popular understandings that have become the stuff of business theory and self-help wisdom as well as daytime talk shows, cable TV programs and a burgeoning therapeutic cottage industry and subculture. Typically, the new happiness discourse espouses a view of emotional life filtered through the lens of economic thought, as in the influential works of Richard Layard whose colorful global surveys of the happiness levels of countries across the world pique the curiosity of the most casual reader (see Layard, 2005). Indeed, Layard's findings have proven influential not just to a lay readership, but at the highest levels of government in some countries, influencing policy discussions in Britain, the United States and Australia (Oishi, 2014).

More precisely, it is a specific and unique formation of economic thought that inscribes the discourse on happiness with its distinctive logic and gives it its singular, penetrating character. This is a contemporary discourse on the economic that makes broad claims for the implicitly opportunistic character of social, personal and emotional existence as a unique enterprise – a neoliberal thought that has become increasingly hegemonic in civic and public discourse, as well as in private and interpersonal life, while an older tradition of economic and social thought rooted in Keynesian welfarism has waned. The story of this shift has become the focus of much recent critical writing. Once, political and economic discourse projected an overarching faith in an implicit human collectivism and in the capacity of states to manage social provisioning, regulate markets and collectivize social risks under economies centrally planned around the shared needs for trust, reciprocity and mutuality. But today, it is the need to foster the freedom of economic actors from these very collective forms, to incentivize enterprising conduct and to responsibilize individual economic risk-taking that forms the nexus of governmental policy (Harvey, 2005). Wendy Larner writes:

> Whereas under Keynesian welfarism, the state provision of goods and services to a national population was understood as a means of ensuring social well-being, neo-liberalism is associated with the preference for a minimalist state. Markets are understood to be a better way of organizing economic activity because they are associated with competition, economic efficiency and choice.
>
> (2000: 5)

Yet there is within the logic of neoliberal government a specific and operative incompleteness, the quality of a problem or a problematization that is central to its functioning. This incompleteness is captured in recent critical work on governmentality studies, and specifically in the concern with neoliberal governmentality, as a set of discursive and institutional practices centered on the shaping of special kinds of subjectivity (Foucault, 1991; Rose *et al.*, 2006).

The governmentality approach applied to the practice of neoliberalism is one that cuts across distinctions between ideology and policy to uncover the political rationalities that operate within each field, and specifically the ways in which these rationalities translate into particular practices for the self-government of neoliberal subjects (McNay, 2009). Neoliberal policies typically involve the restriction of state provisions through budgetary measures designed to give subjects no choice but to adopt enterprising methods, imposing a view of the social field etched in the image of a market abundant with resources, opportunities for mutually beneficial exchanges and competitive advantage, realizable through enterprises of calculation and investment, incentivization, responsibilization, privatization and marketization (McNay, 2009). All of this signifies a process of induced vitality through the self-limitation of a government that operates only indirectly and at considerable distance from its intended objects. The effect is one of excitation and empowerment of subjects through the removal of the constraints imposed by hierarchical institutions and the social commitments they claim to represent. Neoliberalism is, by this token, a fundamentally productive power: it "makes live" by drawing individuals into the competitive production and maximization of their own unique attributes (Foucault, 2007, 2008). Practices of neoliberal governmentality extend these interventionist strategies into the social field but also into the very domain of subjectivity itself where, as Graham Burchell has put it: "Neoliberalism seeks in its own ways the integration of the self-conduct of the governed into the practices of their government and the promotion of correspondingly appropriate forms of techniques of the self" (1996: 29–30).

Neoliberal governmentality thus defines a problem-space for distinct modes of experimentation and intervention wherein society is undone, transformed in the image of the market, and what Burchell (1996: 27) terms an "artificial competitive game" is imposed through the planned minimization of any collectivist alternative to individual competition. The net effect of this is the activation of a distinct range of human potentials and possibilities – the production of a certain neoliberal subjectivity (Burchell, 1996). Indeed, the worst consequence of the welfare state's constraining of the possibilities for individual enterprise is its failure to enable the realization of vital potentials among those it governs – potentials for qualitative differentiation among a populace through the competitive pursuit of opportunities realizable in the terrain of the unfettered marketplace. But for the subject capable of extracting himself from such dependencies (and, conversely, of extracting such inclinations to dependency from himself), the reward comes with the freedom to undertake life as an enterprising endeavor, to take up his own self-cultivation as an enterprising program and, therefore, to invest in himself as would an entrepreneur – on the basis of calculations of investment and return.

In this way, happiness is neoliberal. There is an underlying economic logic which runs through the government of happiness that resonates with the world-view of neoliberal economics and disseminates languages and frameworks mandating a program of enterprising self-government. This is a relation to the self, centered on the stripping away of inherited interdependencies and embedded habits formed around mutuality, and the excitation of a previously suppressed spirit for opportunistic action and entrepreneurship. The current discourse on happiness serves as such a framework by which individuals undertake to problematize aspects of their own conduct, to expunge inherited dependencies in order to optimize personal autonomy and capacity for self-interested initiative. Dependence on the supervision of experts, the propensity to thought-lessly adhere to institutional protocols, a tendency toward idleness or docility, reliance on habitual behaviors shaped in consort with patterned collective life, overinvestment in the judgment of others, or a predisposition to conceive responsibility in collective terms – all are regarded as problematic and cumber-some, as a retardation of the spirit for life, as a result of the overextension of some other vast regime of (welfarist, social) government and, therefore, as an obstruction to the voluntaristic, self-interested, enterprising conduct that is the wellspring of (neoliberal) happiness itself. Indeed, the economism of happiness lies in the very negation of the dependent, constraining and docile attitude that is the legacy of welfare. These elements come together in a new way of talking and writing about human emotional life that has raised the psycho-medical problem of happiness to that of a legitimate scientific object – positive psychology.

With positive psychology, personal happiness has achieved the highest level of transparency and plasticity as an object of positive science, clinical inter-vention and therapeutic manipulation (Gable and Haidt, 2005). Following the publication in 2000 of Martin Seligman's *Authentic Happiness: Using the New Positive Psychology to Realize your Potential for Lasting Fulfillment*, positive psychology has mushroomed into a multibillion-dollar research field and influential self-help discourse, infusing the prefix "positive" to everything from couples therapy, education and marketing to law enforcement and corrections (Seligman, 2000, 2012). In each of these scenarios, the new "positive" psychology is registered as the active, agentive and enterprising counterpart to what it considers traditional psychology, ensconced as it is in the negativity of the disease model, endless reflection on past relations with others and all that makes life a scene of suffering. In the case of positive psychology life coaching, for example, the vocation of the psychotherapist, who mollifies sadness and suffer-ing through patient listening and probing questions, is scorned for stagnating emotional life in the mire of remote and indistinct psychic traumas and heavy-handed expert intervention (Binkley, 2011). In her place, the semi-professional coach engages the patient not so much through a diagnosis of past traumas as through an inspiring reflection on the future as a scene of happiness and self-designed life goals (Brock, 2008). In this regard, positive psychology provides the general theory for a set of interventions into the dynamic of emotional life with the aim of optimizing, improving and developing the happy subject for a life modeled on competition.

Positive psychology's looped resourcing

Positive psychology is a realm of expertise that has achieved broad profes-sional acceptance in academic, public policy and business circles and which, in the space of the past decade, has left a deep imprint on a range of popular therapeutic fields. Aiming to surpass the traditional preoccupation of the psychological professions with negative states (neuroses, psychoses, disorders of various kinds), positive psychology maps out, with the same measure of scientific precision applied to mental pathologies, the psychological states identified with joy, flourishing, expressive well-being and happiness itself. It is possible to date the origin of positive psychology to 1997 when Martin Seligman, renowned for his work on depression and adaptive behavior and recently elected to the presidency of the American Psychological Association (APA), joined forces with Mihaly Csikszentmihalyi, noted psychologist and originator of the concept of "flow," the state of contemplative immersion one attains in an all-consuming activity (Ruark, 2009). Both sought to redress the traditional preoccupation of American psychology with familiar problems of disease, pathology and mental illness through a novel research agenda concentrated on those conditions that make individuals thrive. With the intent of overcoming the vagaries and methodological flimsiness that had hampered previous efforts to treat the positive potentials of human well-being (particularly those identified with the humanistic psychology of Maslow and Rogers), happiness, the two argued, could now be measured objectively and scientifically through empirical clinical research and controlled through precise therapeutic techniques. Buoyed by their conversations, Seligman resolved to make positive psychology the theme for his tenure as president of the APA, and within a few years, the field had exploded.

Since the 2000 publication of Seligman's bestselling work *Authentic Happiness* – the undisputed Holy Writ of this expanding field – the new discourse on happiness has developed into a dynamic cultural phenomenon, earning repute both within academic psychology and in a variety of applied fields from business and public policy to the heady world of self-help publishing. The creation of the John Templeton Prizes in Positive Psychology, two special issues of the *American Psychologist*, a number of handbooks devoted to the topic, several summits and a major international conference all occurred within five years of the initial conversations between the field's founders. And in the time since the publication of Seligman's book, positive psychology has consolidated its hold on academic psychology. Competitive programs in positive psychology have been established at the University of Pennsylvania, Harvard University and the University of East London; Csikszentmihalyi himself has founded a new PhD program in positive psychology at Claremont University and course offerings in positive psychology have become the norm in leading departments worldwide. Financial support for research has also grown rapidly. In addition to recent infusions of support from the National Science Foundation and the US Department of Education, funding in excess of $226 million has been provided to positive psychology researchers by the National Institute of Mental Health (Ruark, 2009; Wallis, 2005). In addition to the $200,000 prizes it has awarded annually since 2000 for new research in positive psychology, The John Templeton Foundation recently offered Seligman a grant of $6 million to encourage collaborative research across the fields of positive psychology and neuroscience.

The new discourse on happiness has influenced a range of institutional, managerial and policy conversations, variously centered on the government of individuals, communities and organizations through appeals to their capacity to feel good about their situations by perceiving them positively. Happiness results from the cognitive outlooks of individuals; to the extent that people can be brought to assess their situations and themselves in a favorable light, the resulting emotional flush will move them to perform on such a superior level as to produce results that actually confirm this initial positive assessment. The task, then, is to create the conditions, or to teach the specific techniques, through which circumstantial optimism and appreciative self-regard can be intentionally cultivated by individuals within their own outlooks. Significantly, this is not undertaken through a treatment regimen, counseling or any therapeutic practice requiring the supervision of an institutional expert of any kind. The cultivation of a positive outlook is the handiwork of any organizational director (teachers, human resource managers, workplace counsellors) who inspires the self-motivated individual to undertake a set of exercises and interventions into his own mundane thought processes. One example of an institutional application of positive psychology is that of "positive education" – developed by Seligman at the Center for Positive Psychology at the University of Pennsylvania – which has been adopted by schools in the United States, Britain and Australia (Waite, 2007; Seligman *et al.*, 2009). Rather than castigating students for their weaknesses and flaws, the curriculum asks students to identify their unique strengths and assets, and includes specific methods by which students might cultivate and sustain this self-regard in their own lives, such as a lesson that concludes with end-of-the-day gratitude reflections designed to enhance positive outlooks (Chen and McNamee, 2011).

In a similar spirit, business has welcomed positive psychology and incorporated its appreciative regard for the positive functions of organizations and enterprises as tools for management. In 2002, the business school at the University of Michigan created a program in Positive Organizational Scholarship, and in 2004, Case Western Reserve University opened a similar program in Positive Organizational Development. Business leaders are taught to view the potentials and assets of organizations and their staffs while imparting to workers small techniques for the enhancement of such appreciative outlooks, woven into the patterns of their daily rounds. These range from keeping records of their own and others' professional accomplishments to the ritual acknowledgment at the start of staff meetings of organizational successes and strengths. Graduates from these programs have brought the assets of positive psychology to firms such as Ann Taylor Stores and Toyota Motor Corporation (Hamburg Coplan, 2009; Linley and Harrington, 2010). Even the US military has incorporated positive psychology methods into its basic training courses, instructing soldiers to direct their thoughts to positive interpretations of events when, for example, a call is placed from the battlefield to one's spouse, who appears to be away from home on a weekend or evening – she's not having an affair; she's working late or gone shopping. In short, happiness is a resource with unlimited organizational value, a link between the present and the future and, therefore, worth cultivating in the emotional dispositions of students, soldiers, workers, prisoners and spouses, and in the general population (Harms *et al.*, 2013).

Perhaps most impressive, however, is the success of positive psychology as a popular cultural and media phenomenon. Regional and national happiness rankings have proven eye-grabbing media fare for readers and viewers world-wide, and a 2005 *Time Magazine* cover story on positive psychology, declaring it the "science of happiness," expanded public curiosity on this phenomenon (Wallis, 2005). Professor Tal Ben-Shahar's positive psychology class (from which he developed materials for his 2007 bestselling book *Happier: Learn the Secrets of Daily Joy and Lasting Fulfillment*) was, for a time, publicly celebrated as the most popular class at Harvard University. And on the self-help shelves, dozens of titles brandishing the scientific credentials of the new psychology strive to set themselves apart from the mushier offerings of self-help and new age gurus. A cover story in *Psychology Today* reports that while only 50 new popular nonfiction titles addressed the topic of happiness in 2000, that number had grown to 4,000 by 2008 (Flora, 2009).

Conceptually, the core elements of positive psychology are relatively easy to grasp, owing to the field's penchant for the popular psychology genre. Draw-ing on the legacy of humanistic psychology, positive psychologists refute the pessimism of the "adaptive" tradition and focus on the life-affirming poten-tials, energies and vital forces residing within the individual psyche. Carl Rogers, Abraham Maslow and proponents of the movement for self-realization in the 1960s and 1970s had argued for the need to evolve a therapeutic metho-dology and a style of interpersonal life that transcends the self-recrimination imposed on the individual by demanding social norms while accepting uncon-ditionally the qualities and character of individuals in a spirit of warmth and affirmation – what became known as client-centered psychotherapy (Froh, 2004). Positive psychology is similar in its optimistic portrayal of happiness as a radiant personal potential and its emphasis on the need to overcome negative self-assessments, although in this case, the therapeutic task is radically disengaged from relations with others and turned over to the individual himself. The happy subject is taught to maximize happy emotions through the direct manipulation of his own thoughts, understood as resources for the optimiza-tion of an emotional state – a characteristic that positive psychology inherits from its other great forebear, cognitive behavioral psychology. Cognitivist approaches typically reverse the old Freudian axiom that thoughts are the expression of underlying emotional dynamics, which are themselves rooted in psychobiographical experiences. Instead, everyday thoughts are understood to determine emotional states; and where these thoughts can be directly manipulated by sheer acts of will (making oneself think about this or that), it follows that happiness can be produced by consciously directing one's thoughts to happy subjects with the same intentionality one might pursue in a fitness regime. Positive psychologists provide reams of advice on how this is to be done. Through thought interventions, one learns to switch off negative patterns of thinking. These involve planned disruptions of routine mental habits, which forestall the cyclical downward spiral to adaptive emotional states that embed us in the rhythms of daily life. Indeed, together with new clinical methodologies for the specific measurement of emotional conditions, wide authority is granted to the individual for the adjustment and manipulation of a static condition – one's happiness, whose intensity can be determined numerically from moment to moment and by the simple and direct method of self-reporting – through the control of one's thoughts (Seligman and Csikszentmihalyi, 2000).

Moreover, positive psychology proposes specific methods for the enhancement not just of states of positive feeling in real life (hedonic pleasure) but also the deeper forms of happiness that derive from the exercise of our chief potentials and unique gifts as individuals (eudaimonic happiness). This kind of happiness, termed "authentic happiness" by Seligman, occurs when a particular set of psychological strengths and virtues unique to each individual are mobilized and put into operation in everyday activities – qualities such as courage, conviction and open-mindedness, whose development through practice in everyday life induces positive self-regard and thus happier emotional states (Seligman, 2000). Seligman recounts the process by which these qualities were arrived at in the development of positive psychology. Together with a colleague, he combed through the "basic writings of all the major religious and philosophical traditions … Aristotle, Plato, Aquinas, Augustine, the Old Testament, the Talmud, Confucius, Buddha, Lao-Tze, Bushido, the Koran, Benjamin Franklin" (2000: 132) to track the recurrence of distinctive positive traits. What emerged was a list of universally held "signature strengths," which include Wisdom, Knowledge, Courage, Humanity, Justice, Temperance and Transcendence. Peterson and Seligman (2004) went on to catalog these qualities in the *Character Strengths and Virtues* handbook, or CSV, proposed as positive psychology's counterpart to the inventory of pathological states numbered in the *Diagnostic and Statistical Manual of Mental Disorders* (DSM) (see Maddux, 2002).

At the foundation of positive psychology, then, is a deep belief in the incompleteness of the project of happiness itself, in the plasticity of emotional states, and in the opportunistic conduct of the happy subject as one susceptible to the suggestive power of optimistic and pessimistic thought. Negative emotional states derive from the perception of one's own helplessness to make oneself happy, the inability to transcend one's routines or an overdependence on the emotional patterns that develop from unexamined, shared, social life. Positive emotions, on the other hand, come with the embrace of one's power to change one's emotional well-being and with the assumption of responsibility for those emotions. In the first case, one is unhappy and believes that one cannot act to make oneself happy because one is too rooted in a way of life and in a set of dependencies on others, which makes one even more unhappy. In the second, one brings oneself to see that one can escape the limits imposed by a socially embedded life by viewing people and situations not as obligations or as externalizations of one's own psychic predicament but as resources for manipulation and optimization. This realization gives one a sensation of emotional exhilaration and forces a cognitive shift, which itself motivates action and brings about the very happier reality one had convinced oneself to believe in initially. Unhappiness is synonymous with the inability to act on one's own, originating in one's acceptance of habitualized outlooks derived from others and tinged by inevitability. To the extent that one realizes that one can make oneself happy through one's own actions, one becomes happy. The practical logic of the happiness discourse is one that invokes the circular effect of a causal loop, or a looped resourcing, in which agency, enterprise and responsibility for oneself are both the means for achieving happiness, and the very content of happiness itself. Through this loop, the apprehension that happiness is within one's reach is a perception that is realized through the taking of actions toward happiness. And by extension, the spiral of docility, resignation, dependence and the reluctance to see the world in ways that break away from the pack and therefore enable one to act on one's own signals not only the absence of happiness but the inhibition and retardation of the potential for happiness – the vital, enterprising life-spirit that is the wellspring of life's activity, or freedom. Thus, the parallel between positive psychology and neoliberal economic thought is clear: the docility of social dependence and the negative thoughts that lull us into states of torpor must be actively uprooted and transformed through an infusion of affirming optimism and the diminishment of any collectivist alternative to enterprise. Dependence, immobility and stasis are anathema to the happiness project. Psychologist Barbara Fredrickson conveys this anxiety around the inertia of dependence this way:

> Gratuitous negativity can hold you hostage, as if you had cinder blocks tied to your ankles and a black hood pulled over your face. It can keep you so constrained and smothered so that you are simply unable to flourish. But the good news is that you have what it takes to free yourself.
>
> (2009: 159)

The looped resourcing of happiness demands that the happy subject train her efforts on these obstructive objects (cinder blocks, black hoods, negativity itself) that suppress the agency and freedom that makes happiness possible. And these things are found in the thoughts and habits that embed the individual in the mutualities that constitute patterned social life.

Such is the productive effect of the discourse on happiness whereby happiness is that emotional medium through which the freedom of the entrepreneurial subject is constituted. Clearly, this is a subject on the move. It is one of enterprise, opportunity and agency. What is less evident, however, is the extent to which this is also a subject of vulnerability and anxiety, one who also seeks in a technology of emotional life a respite from insecurity and fear. Thus, happiness can also be understood in terms of its capacity to speak to the need for security and safety, to provide resilience in the context of instability and risk.

Risk, resilience and emotional life

Anxieties about the future have always plagued human life, and human societies have always mobilized against these uncertainties through collective practices of calculation, anticipation and risk minimization and through the adoption of coordinated, anticipatory, future-oriented dispositions. But today, something truly unique is happening to the way that we respond to risk. Today we have amassed such an immense technology for the assessment and minimization of risks that the apparatus of risk assessment itself has come to operate through a double effect; the control of uncertainty on a societal level through policy and practice dovetails with the production of unique subjects possessing a certain temporal disposition centered on the anticipation of uncertain outcomes. Where discourses of financial forecasting, insurance, epidemiology, genetics, public health, environmentalism and criminology variously predict the probabilities of economic recession, disease and street violence, such discourses ultimately impose themselves upon the daily attitudes and self-understandings of lay individuals, who adopt and incorporate such views into their own outlooks as subjects. We have become citizens of the risk society, ever more anxious about the future, ever more ready to measure and plan our futures on the basis of an increasingly sophisticated calculus of probability (Beck, 1992; Giddens, 1992; Beck *et al.*, 1994).

Today, with a vast expansion in the range of technologies for the identification and measurement of more and more subtle gradations of risk, the attribution of rationalist notions of causality and effect has become subverted by the sheer ubiquity and mobility of societal risk itself. This effect was coupled with the curtailed influence, under the ever-expanding pressures of the market, of those collective institutions (classes, states, civic institutions, labor unions, communities) whose function it was to implement risk-minimizing policies. While under the conditions of an earlier industrial modernity, it might have been possible for a corporate body of experts to specify and delineate these risks spatially and temporally, today the responsibility for calculation and preparedness has been increasingly shifted to individuals themselves. This is particularly true with regard to social inequality under the conditions of reflexive modernity, which becomes redefined in terms of an individualization of social risks, or as Ulrich Beck puts it: "The result is that social problems are increasingly perceived in terms of psychological dispositions: as personal inadequacies, guilt feelings, anxieties, conflicts and neuroses" (1992: 100).

For this self-reflexive person, the insufficiently developed risk awareness exhibited in mundane life appears as an open-ended problem, framed by expert discourses and available special knowledges, or as the objects of personal innovation and enterprising action (Dean, 1999: 176–97). The production of identity under these conditions entails the projection into the future of a trajectory of expected outcomes and events. In an effort to direct this identity and control these events, the individual is left scrambling for resources; bereft of the supports once provided by institutions such as the state, family, religious institutions and civil society, temporalized identities are projected into a world of uncertainty and risk against which the individual has only her own resources to draw upon. The individualization of risk, therefore, orients the individual's conduct toward the eventuality of such risks, expanding temporal horizons and "responsibilizing" conduct itself through the assumption of a calculating, anticipatory regard for the future even as it exacerbates the anxieties and uncertainties that accompany the anticipation of incalculable liabilities.

A theory of the risk society provides an account of the temporalization of subjectivity that shares many features with that attributed to the happiness discourse; changing social arrangements and an expanding apparatus of government affect changes in the ways subjects think about and experience time. The differences come with the manner in which, within the discourse on happiness, risk technologies and the future orientations they inscribe are assimilated into personal life as a generative, constitutive force. Within that constellation of research I have earlier referred to as governmentality theory, a somewhat nominalist theoretical perspective is taken, not on the general form of a specific phase of modernity but on the specific institutions, technologies and apparatuses that affect the temporalization of subjectivity. A contemporary technology for the government of risk is traced to a specific cast of agents of governmental rationality: health researchers, demographers, financial forecasters, statisticians and environmental experts, coupled with the practical advice of therapists, consultants, legal advisors, insurance agents, real estate brokers and investment bankers, all of which combine to form a matrix of authorities and discourses on risk, and whose rationalities of risk management are transposed, through the mechanism of governmentality, onto the outlooks and conducts of subjects themselves. For example, insurers, François Ewald (1991) has written, render the vagaries and uncertainties of fate in a calculable form reducible to a set of capital equivalencies. They establish corresponding investments required in the present to secure capital compensations should such improbable outcomes occur (Ewald, 1991; Dean, 1999: 183–8). And, importantly, insurantial risk technologies induce the individual into a sustained form of life through the adoption of practices oriented toward future risks, insured or not. "Insurance is a moral technology," writes Ewald, and:

> To calculate a risk is to master time, to discipline the future. To conduct one's life in the manner of an enterprise indeed begins in the eighteenth century to be a definition of a morality whose cardinal virtue is providence. To provide for the future does not just mean not living from day to day and arming oneself against ill fortune, but also mathematizing one's commitments.
>
> (1991: 207)

Indeed, such a moral technology of insurance constitutes an important element of daily conduct, not only under earlier stages of societal modernization but also under the present conditions of risk in advanced capitalist societies.

Mitchell Dean, following Pat O'Malley, has termed this everyday conduct the "new prudentialism," which entails

> the multiple responsibilization of individuals, families, households and communities for their own risks – of physical and mental ill-health, of unemployment, of poverty in old age, of poor educational performance, of becoming victims of crime. Competition between public (state) schools, private health insurance and superannuation schemes, community policing and "neighbourhood watch" schemes, and so on, are all instances of contriving practices of liberty in which the responsibilities for risk minimization become a feature of the choices that are made by individuals, households and communities as consumers, clients and users of services.
>
> (1999: 166; see also O'Malley, 1996)

Today, emotional well-being could well be included among those "multiple responsibilizations" for which the enterprising risk subject is implicated. Where events that might bring on a state of sadness or depression loom unpredictably in the future, it is the responsibility of the happy subject to plan for their possibility as well as to become proactive in positively producing the capacities within oneself to deal with emotional liabilities, to assess and plan for the exigencies of future emotional states by enriching and fortifying one's happiness in the present, and taking a productive stance in the preservation and maximization of personal well-being. Positive psychology has responded to the problem of emotional risk management through its incorporation of a theory of psychological "resilience," a term and a field of research developed in the 1990s to investigate the capacity for endurance exhibited by certain individuals under physically or emotionally adverse conditions. Resilience is "the process of, capacity for, or outcome of successful adaptation despite challenging or threatening circumstances" (Masten *et al.*, 1990: 426). The identification of resilient behavior involves the singling out of the various attributes of character that have enabled individuals to show resilience in the face of challenge, most import being positive affect, or a long sustained state of happiness prior to the adverse events themselves.

A concern with resilience was popularized to a general self-help readership and introduced as a concern for positive psychology by Karen Reivich and Andrew Shatté's 2003 bestseller *The Resilience Factor: 7 Keys to Finding Your Inner Strength and Overcoming Life's Hurdles*. For these authors, the production of emotional assets in anticipation of traumatic events in an uncertain future is accomplished through a program of emotional consolidation with deep roots in cognitivist theory. Indeed, resilience researchers' emphasis more generally on the cultivation of a future orientation as a specific skill exhibits clear resonances of Snyder's theory of hope and Bandura's self-efficacy; one must retool one's sense of time and instill within one's cognitive disposition a unique regard for potential hazards on the horizon if one is to successfully confront the problems of the future. Resilience skills enable patients to "bounce back from the adversities – recovering from addiction, coping with bereavement, dealing with job loss or divorce – that so often lead to clinical depression and anxiety" (Reivich and Shatté, 2003: 11). More precisely, the book outlines several lessons the reader will undergo on the path to enhanced resilience, each bringing with it a specific resource in the coming confrontation with emotional risk. These lessons are identified: (1) learn your ABCs – "We'll teach you to 'listen' to your thoughts, to identify what you say to yourself when faced with a challenge, and to understand how your thoughts affect your feelings and behavior"; (2) avoid thinking traps – "We'll teach you to identify the ones you habitually make and how to correct them"; (3) detect icebergs – "We'll teach you how to identify your deep beliefs and determine when they are working for you and when they are working against you"; (4) challenge beliefs – "We'll teach you how to test the accuracy of your beliefs about problems and how to find solutions that work"; (5) put it in perspective – "We'll teach you how to stop the what-ifs so that you're better prepared to deal with problems that really do exist or are most likely to occur"; (6) be calm and focus – "We'll show you how to stay calm and focused when you're over-whelmed by emotion or stress so you can concentrate on the task at hand"; (7) real-time resilience – "We'll teach you a powerful skill so that you can quickly change your counterproductive thoughts into more resilient ones" (Reivich and Shatté, 2003: 13–14). And in the field of positive psychology, the need to forcibly generate an optimistic regard as a defense against future emotional traumas is more explicitly argued by Seligman himself. In *Learned Optimism: How to Change Your Mind and Your Life*, Seligman describes the results of his efforts to prevent depression among schoolchildren through the teaching of specific techniques for the cultivation of positive future orientation, or optimism:

> Optimists do much better in school and college, at work and on the playing field. They regularly exceed the predictions of aptitude tests. When optimists run for office, they are more apt to be elected than pessi-mists are. Their health is unusually good, they age well, much freer than most of us from the usual physical ills of middle age. Evidence suggests they may even live longer.
>
> (2006: 5)

In short, the optimistic regard for the future is the figure of happiness; yet it is a happiness that only comes about as the result of a specific effort and as the result of a task of self-optimization. Optimism is not a natural attribute, but must be actively mobilized within one's outlooks through the negation of optimism's logical opposite and necessary antithesis – pessimism. Pessimism, Seligman argues, is also a relation to the future. It is a deeply embedded habit of thought that inevitably anticipates the worst and forfeits any possibility of intervening in this outcome to the overwhelming forces of fate. "At the core of the phenomenon of pessimism is another phenomenon—that of helplessness. Helplessness is the state of affairs in which nothing you choose to do affects what happens to you" (Seligman, 2006: 5). Futurelessness and dependency carve deep grooves in one's mental routines, into which our thoughts, expectations and reflections are inevitably drawn and where they are condemned to circle endlessly. Indeed, pessimism suffers from a specific temporal sensibility that stifles a proactive future orientation and must, for this reason, be uprooted and transformed – a project that could potentially extend to the multitudes of people who, unbeknownst to them, carry some buried trace of this disabling mental and emotional state. Seligman writes:

> I have learned that it is not always easy to know if you are a pessimist, and that far more people than realize it are living in this shadow. Tests reveal traces of pessimism in the speech of people who would never think of themselves as pessimists; they also show that these traces are sensed by others, who react negatively to the speakers.
>
> A pessimistic attitude may seem so deeply rooted as to be permanent. I have found, however, that pessimism is escapable.
>
> (2006: 5)

Taken together, the regulation of happiness serves these differentiated, though complimentary, functions. It is at once a resource to be mobilized and applied opportunistically to varied situations and a defense against uncertainty for a subject made vulnerable under circumstances of social disembeddedness and risk. However, as contradictory as these tendencies may seem, what is important to grasp is the dynamic relation they establish in which risk exerts a specifically vitalizing effect on subjectivity itself. As Julian Reid writes, "[t]he neoliberal subject is not a subject which can conceive the possibility of securing itself from its dangers, but one which believes in the necessity of life as a permanent struggle of adaptation to dangers" (2012: 149). The happy subject is not one that clearly differentiates opportunity from danger or vulnerability from well-being. Indeed, happiness, as it is articulated in this new discourse, folds the two into each other, whatever the emotional toll.

Note

1 I am grateful for having had the opportunity to present a keynote address at the Behaviour Change and Psychological Governance, ESRC Seminar Series, at the University of Birmingham in June of 2014. This address was titled "Happiness as Enterprise," and provided a summation of key themes from my recent book *Happiness as Enterprise: An Essay on Neoliberal Life* (SUNY, 2014). For the purposes of this volume, I have derived the present article principally from select passages of that book, composed here so as to highlight issues central to the focus of the conference, particularly the theme of psychological resilience.

References

Ahmed, S. (2008) *The Promise of Happiness*. Duke University Press, Durham, NC.
Beck, U. (1992) *Risk Society: Towards a New Modernity*. Sage, London.

Beck, U., Giddens, A. and Lash, S. (1994) *Reflexive Modernization. Politics, Tradition and Aesthetics in the Modern Social Order.* Stanford University Press, Stanford, CA.

Ben-Shahar, T. (2007) *Happier: Learn the Secrets to Daily Joy and Lasting Fulfillment.* McGraw-Hill, New York.

Binkley, S. (2011) "Psychological life as enterprise: social practice and the government of neoliberal interiority." *Journal of the History of Human Sciences*, 24(3): 83–102.

Binkley, S. (2014) *Happiness as Enterprise: An Essay on Neoliberal Life.* SUNY Press, Albany, NY.

Brock, V. (2008) Grounded Theory of the Roots and Emergence of Coaching. PhD thesis, International University of Professional Studies, Maui.

Burchell, G. (1996) "Liberal government and techniques of the self," in A. Barry, T. Osborne and N. Rose (Eds.) *Foucault and Political Reason.* UCL Press, London, pp. 19–36.

Chen, J. and McNamee, G. (2011) "Positive approaches to learning in the context of preschool classroom activities." *Early Childhood Education Journal*, 39(1): 71–78.

Cruikshank, B. (1999) *The Will to Empower: Democratic Citizens and Other Subjects.* Cornell University Press, Ithaca, NY.

Dean, M. (1999) *Governmentality: Power and Rule in Modern Society.* Sage, London.

Evans, B. and Reid, J. (2014) *Resilient Life: The Art of Living Dangerously.* Polity, New York.

Ewald, F. (1991) "Insurance and Risk," in G. Burchell, C. Gordon and P. Miller (Eds.) *The Foucault Effect: Studies in Governmentality*, University of Chicago Press, Chicago, IL, pp. 197–210.

Flora, C. (2009) "The pursuit of happiness" [online]. *Psychology Today*, 1 January. Available at: https://www.psychologytoday.com/articles/200901/the-pursuit-happiness

Foucault, M. (1991) "Governmentality," in G. Burchell, C. Gordon and P. Miller (Eds.) *The Foucault Effect: Studies in Governmentality.* University of Chicago Press, Chicago, IL, pp. 87–104.

Foucault, M. (2007) *Security, Territory, Population. Lectures at the Collège de France, 1977–78* (trans. G. Burchell). Palgrave, New York. Foucault, M. (2008) *The Birth of Biopolitics: Lectures at the Collège de France* (trans. G. Burchell). Palgrave, New York.

Fredrickson, B. (2009) *Positivity: Top Notch Research Reveals the 3-to-1 Ratio that Will Change Your Life.* Three Rivers Press, New York.

Froh, J. J. (2004) "The history of positive psychology: truth be told." *NYS Psychologist*, 16(3): 18–20.

Gable, S. and Haidt, J. (2005) "What (and why) is positive psychology?" *Review of General Psychology*, 9(2): 103–110.

Giddens, A. (1992) *The Transformation of Intimacy: Sexuality, Love, and Eroticism in Modern Societies.* Stanford University Press, Stanford, CA.

Hamburg Coplan, J. (2009) "How positive psychology can boost your business" [online]. *Bloomberg Businessweek*, 13 February. Available at: www.bloomberg.com/news/articles/2009-02-12/how-positive-psychology-can-boost-your-business

Harms, P. D.Herian, M. N., Krasikova, D. V., Vanhove, A. and Lester, P. B. (2013) The Comprehensive Soldier and Family Fitness Program Evaluation Report #4: Evaluation of Resilience Training and Mental and Behavioral Health Outcomes. Positive Psychology Center, University of Pennsylvania, Pennsylvania, PA. Available at: https://ppc.sas.upenn.edu/sites/ppc.sas.upenn.edu/files/csftechreport4mrt.pdf (accessed 8 December 2015).

Harvey, D. (2005) *A Brief History of Neoliberalism*. Oxford University Press, Oxford.

Holmes, M. (2010) "The emotionalization of reflexivity." *Sociology*, 44(1): 139–154.

Holmes, M. (2015) "Researching emotional reflexivity." *Emotion Review*, 7(1): 61–66.

Illouz, E. (2007) *Cold Intimacies*. Polity, New York.

Larner, W. (2000) "Neo-liberalism, policy, ideology, governmentality." *Studies in Political Economy*, 63(1): 5–25.

Layard, R. (2005) *Happiness: Lessons from a New Science*. Penguin, New York.

Linley, A. P. and Harrington, S. (2010) *Oxford Handbook of Positive Psychology and Work*. Oxford University Press, Oxford.

McNay, L. (2009) "Self as enterprise: dilemmas of control and resistance in Foucault's the birth of biopolitics." *Theory, Culture and Society*, 26(6): 55–77.

Maddux, J. (2002) "Stopping the 'madness': positive psychology and the deconstruction of the illness ideology and the DSM," in C. R. Snyder and S. J. Lopez (Eds.) *Handbook of Positive Psychology*. Oxford University Press, New York, pp. 13–25.

Masten, A. S., Best, K. M. and Garmezy, N. (1990) "Resilience and development: contributions from the study of children who overcome adversity." *Development and Psychopathology*, 2(4): 425–444. Oishi, S. (2014) "Can and should happiness be a policy goal?" *Policy Insights from the Behavioral and Brain Sciences*, 1(1): 195–203.

O'Malley, P. (1996) "Risk and responsibility," in A. Barry, T. Osborne and N. Rose (Eds.) *Foucault and Political Reason: Liberalism, Neo-liberalism, and Rationalities of Government*. UCL Press, London, pp. 189–208.

Peterson, C. and Seligman, M. (2004) *Character Strengths and Virtues: A Handbook and Classification*. American Psychological Association, Washington, DC.

Reid, J. (2012) "The Neoliberal Subject: resilience and the art of living dangerously." *Revista Pléyade*, 10 (July–December): 143–165.

Reivich, K. and Shatté, A. (2003) *The Resilience Factor: 7 Keys to Finding Your Inner Strength and Overcoming Life's Hurdles*. Broadway Books, New York.

Rose, N., O'Malley, P. and Valverde, M. (2006) "Governmentality." *Annual Review of Law and Social Science*, 2: 83–104.

Ruark, J. (2009) "An intellectual movement for the masses: 10 years after its founding, positive psychology struggles with its own success" [online]. *Chronicle of Higher Education*, 3 August. Available at: http://chronicle.com/article/An-Intellectual-Movement-fo/47500/ (accessed July 2010).

Seligman, M. (2000) *Authentic Happiness: Using the New Positive Psychology to Realize your Potential for Lasting Fulfillment*. Free Press, New York.

Seligman, M. (2006) *Learned Optimism: How to Change Your Mind and Your Life*. Vintage Books, New York.

Seligman, M. (2012) *Flourish: A Visionary New Understanding of Happiness and Well-being*. Free Press, New York.

Seligman, M. and Csikszentmihalyi, M. (2000) "Positive psychology: an introduction." *American Psychologist*, 55(1): 5–14.

Seligman, M., Ernst, R. M., Gillham, J., Reivich, K. and Linkins, M. (2009) "Positive education: positive psychology and classroom interventions." *Oxford Review of Education*, 35(3): 293–311.

Waite, R. (2007) "Happiness classes 'depress pupils'" [online]. *The Sunday Times*, 9 September. Available at: www.timesonline.co.uk/tol/news/uk/article2414778.ece (accessed July 2010).

Wallis, C. (2005) "The new science of happiness." *Time Magazine*, 9 January. Available at: http://content.time.com/time/magazine/article/0,9171,1015832,00.html

Zevnik, L. (2014) *Critical Perspectives in Happiness Research: The Birth of Modern Happiness.* Springer, New York.

4 Therapeutic governance of psycho-emotionally vulnerable citizens

New subjectivities, new experts and new dangers

Kathryn Ecclestone

Introduction

Multiple crises of late capitalism seem increasingly to be expressed through the media, in policy discourses and in everyday life as 'relentlessly repetitive problematisations' (Isin, 2004: 228) or as 'wicked' problems; namely, intractable, highly complex, changing and contested (e.g. Richardson, 2011; Bache *et al.*, 2015). A particular focus is mental illness, presented as a worsening global epidemic by the World Health Organization, UNICEF and the OECD as well as pharmaceutical companies, psychology professional bodies and global corporations (e.g. Mills, 2014; Davies, 2015). In a British context, especially since the election of a Labour government in 1997, these concerns have become intertwined with fears about the general psycho-emotional states of citizens and the difficulty of childhood and life transitions. One effect across the education system and other social policy settings has been to create slippery elisions between wellbeing, mental health, resilience and 'character'.

As part of a therapeutic behaviour change agenda under successive governments over this time period, these elisions have enabled diverse, *ad hoc* ideas and practices from counselling, psychotherapy, neurolinguistic programming, positive psychology, mindfulness, self-help and peer-based therapies to permeate numerous areas of research and practice in education and social policy (Ecclestone, 2013). In exploring the intersection of these developments with wider behaviour change policy agendas, this chapter focuses on the images of human subjects these agendas generate and respond to, the new types of experts emerging to govern them, and the dangers that arise. In setting the scene for discussion of these themes, I outline examples of the political and cultural normalisation of a therapeutic behaviour change agenda.

In educational settings from early years to university, numerous policy reports reflect a wide, influential consensus that an interrelated set of psycho-emotional attributes, dispositions and behaviours associated with emotional regulation/intelligence/literacy, resilience, stoicism, optimism, character, hope, aspiration and community-mindedness can be taught, learned and transferred between contexts and over time as an essential foundation for successful educational and life functioning (e.g. Sharples, 2007; Paterson *et al.*, 2014; O'Donnell *et al.*, 2014).

Between 1998 and 2010, successive governments sponsored myriad generic or universal initiatives seeking to develop this extensive list of psycho-emotional foundations. During the same period, there has been a big increase in targeted activities for those with formal diagnoses of psychological, emotional and behavioural conditions and disorders (e.g. Harwood and Allen, 2014). Although the Conservative-led coalition government (formed in 2010) replaced political sponsorship of national social and emotional learning programmes with renewed interest in the much older notion of 'character', its Conservative successor (elected in 2015) has elided this with an intensification of concern about mental health. Under the banner of character, the government funds various initiatives to promote the skills and dispositions of earlier wellbeing initiatives (see Morgan, 2015a, 2015b and Ecclestone, 2013 for discussion).

Claims about a need for approaches to strengthen psycho-emotional skills also pervade family and parenting policies. For example, since the late 1990s, ideas from neuroscience have been used to embellish successive governments' resurrection of determinist ideas from the 1930s about the lifelong effects of neglect, poor attachment and dysfunctional parenting. All mainstream political parties now agree that the inner states of human subjects determine the relationship between psycho-emotional responses and behaviours in complex, non-linear ways from conception, requiring 'early intervention' – especially for families deemed 'inadequate' and 'troubled' – to create a virtuous circle of engagement, inclusion, aspirations, achievement and social mobility (e.g. Field, 2010; Allen, 2011; Department for Education and WAVE Trust, 2013). For example, proposed antenatal and postnatal care assessments privilege the emotional bond between mother and child and prospective mothers' attachment behaviour, especially for parents considered to have problems (Department for Education and WAVE Trust, 2013). This contemporary turn in notions of 'good parenting' and associated resource allocations continues the shift amongst charities and state agencies since the 1890s from ameliorating the effects of poverty and 'poor hygiene' to intervening in the psycho-emotional dynamics of families from an increasingly early stage (Gillies, 2015; see also Myers, 2010 and Stewart, 2011).

More broadly, the politically and culturally popular idea that psycho-emotional dysfunction generates damaging individual and social legacies together with the wider normalization of associated assessment and intervention are inseparable from what sociologists call 'therapeutic culture'; namely, the prevalence of ideas, practices and assumptions from branches of psychology, therapy, counselling and self-help throughout popular culture, politics, education, legal and welfare systems (see Rose, 1999; Nolan, 1998, Furedi, 2004, Wright, 2011, Ecclestone and Hayes, 2009). In educational settings at all levels, these ideas and practices have become intertwined with long-running criticisms of an overly cognitive approach to learning and the promotion of more attention to affective dimensions. For example, notions such as 'learning to learn', 'being an independent/collaborative learner/reflective practitioner' and 'co-producing learning' are embedded routinely in pedagogies and related assessments such as personal development portfolios, professional journals, mentoring programmes, and staff appraisals and reviews. These normalise regular assessments of dispositions, attitudes and behaviours such as self-esteem, engagement, confidence, resilience, emotional management and motivation. At the same time, lifestyle and popular media, books, articles, and software applications that monitor psycho-emotional states, together with everyday casual speculations amongst colleagues, friends and family, make 'emotional issues', people's ability to deal with situations, and psychological causes of behaviour a constant source of cultural preoccupation.

These political, institutional and cultural dimensions to the normalisation of therapeutic behaviour change approaches across different policy arenas frame my exploration in this chapter of the ways in which vulnerability has become a key tenet of contemporary crisis discourses about the 'ideal' citizen as rational, autonomous, reasoning and choosing, and how it is integral to contemporary forms of psychological governance. The chapter aims to open new theoretical and empirical lines of enquiry in relation to three interrelated cultural and political phenomena: the rise of vulnerable subjects as targets for new forms of therapeutic governance; the role of the state in responding to and producing particular subjectivities; and the dangers arising from the erratic, sometimes intrusive, state-sponsored intervention market that these developments generate.

My exploration of these themes is structured as follows. The first section relates a powerful contemporary sensibility of endemic psycho-emotional vulnerability to long-running philosophical and political debates about the salience and desirability of the archetypal rational, autonomous, choice-making liberal and neo-liberal human subject as a target for governance; it also highlights some historical and contemporary examples of engagement between wider behaviour change policy interests and these subjects. I then use a Weberian understanding of authority to explore how a growing market of therapeutic interventions invokes dichotomies between different images of subjects as targets for governance that, in turn, legitimise new types of psy-expertise. The third section illuminates dangers that arise when governance becomes rooted in cultural anxiety about psycho-emotional vulnerability as a defining feature of everyday experience. I conclude with propositions that the privileging of vulnerable subjects in therapeutic forms of governance reflects the state's growing uncertainty about its ability to govern through notions of the archetypal liberal and neo-liberal subject.

Universalizing vulnerability

The rise of a 'vulnerability zeitgeist'

A significant expansion of official criteria for defining 'the vulnerable' as targets for British social policy reflects what Kate Brown calls a 'vulnerability zeitgeist'. Here, successive British governments between 1998 and 2010 have created diffused, malleable and increasingly expanding criteria that take account of changing rationales, preoccupations and preferences (Brown, 2015; see also McLaughlin, 2011). The Law Commission's 1996 criteria defining a relatively small minority of targeted individuals and groups as vulnerable because they could not manage everyday life independently by 'reason of mental or other disability, of age or illness and who is or may be unable to take care of him or herself, or unable to protect him or herself against significant harm or exploitation' began to expand through ideas about 'social exclusion' and 'those leading chaotic lives' promoted by the 1998–2010 Labour government (Social Exclusion Unit, 1999). Notably, the Labour government's Care Standards Act 2000 expanded official criteria for vulnerability to include anyone receiving prescribed medical, counselling or palliative care (McLaughlin, 2012).

In subsequent expansions, the UK Cabinet Office Social Exclusion Task Force (2007) referred to disadvantages associated with social exclusion: worklessness, poor or overcrowded housing, parents' lack of academic or vocational qualifications, mothers with mental health problems, a family member with long-running disability or illness, income 60 per cent below the median, and inability to afford certain food and clothing items. The Office for Standards in Education (2012) currently elides 'disadvantaged' and 'vulnerable' to encompass migrant children, those with special educational needs, and pupils who are disengaged or who are simply not meeting their targets. Here, notions of support and safeguarding widen vulnerability almost infinitely to children 'whose needs, dispositions, aptitudes or circumstances require particularly perceptive and expert teaching and, in some cases, additional support' (2012: 6).

The overall effect of such diffused understandings is that even vaguer notions such as 'tendencies to vulnerability' and 'multiple vulnerabilities' merge with nebulous ideas of 'needs', 'harms' and 'risks' in official and everyday professional discourses, thereby presenting a widening constituency of citizens as in need of some sort of intervention (e.g. Jopling and Vincent, 2015; Ecclestone and Lewis, 2014).

Universal vulnerability as a political project

Self-defined radical, critical and progressive standpoints challenge what they regard as attempts to use vulnerability in maintaining the status quo or to ameliorate vulnerability as a deficiency or weakness (see McLaughlin, 2011). Hoping to transform vulnerability into something transgressive and therefore politically progressive, a prevailing theme is to recognise it as an attribute of an understanding, empathetic citizenship and a 'universal' dimension of human experience and identity (Beckett quoted by McLeod, 2012: 22). Other ideas about social justice present recognition of and responses to vulnerability as integral to the ethos and actions of an inclusive, responsive, non-neo-liberal state (e.g. Fineman, 2008; see also Ecclestone and Brunila, 2015 for discussion). More widely, vulnerability as an existential fact of death, risk of illness and disability becomes integral to challenging the 'precarity' of late capitalism's dismantling of the 'conditions for living' (e.g. Paur, 2012; Goodley in Ecclestone and Goodley, 2014).

Yet, in parallel to the widening meanings evident in official policy and everyday discourse, vulnerability as part of an alternative political project becomes similarly expansive. It increasingly goes far beyond the idea that physical and psychological vulnerabilities arise from the material, structural demands and inequalities of capitalism to see them as an inherent psychological state generated by social and economic progress *per se*. This encourages citizens and governments to lower material expectations and elevate the dimensions of wellbeing and quality of life (see Frawley, 2014). From various perspectives, vulnerability has become 'a cultural metaphor, a resource drawn upon by a range of parties to characterize individuals and groups and to describe an increasingly diverse array of human experience' (Frawley, 2014: 11). Seen in this light, a predominantly psycho-emotional view contrasts strongly with older political perspectives that saw vulnerability predominantly as a material condition that required proper recognition of people's diverse, sometimes idiosyncratic, yet agentic psychological and material resources for dealing with adversities and the empowerment they often seek and demonstrate in doing so – ultimately requiring material remedies (e.g. Valentine and Skelton, 2003).

Interplays between the rational and vulnerable subject

The normalisation of vulnerability as a universal, mainly psycho-emotional, condition for a more inclusive and emancipatory politics can be seen as a contemporary strand in long-running political and philosophical critiques of the privileging of

> active individuals seeking to 'enterprise themselves', to maximize their quality of life through acts of choice, according their life a meaning and value to the extent that it can be rationalized as the outcomes of choices made or choices to be made.
>
> (Miller and Rose, 2008: 214)

In this vein over the past 50 years or so, various strands of feminist, post-modernist, post-structuralist and post-humanist thought have presented this subject not merely as unrealistic, anachronistic and a political and cultural myth, but also – in its portrayal as a universal aspiration – as an oppressive Western, male, heteronormative, able-bodied 'othering' of anyone outside those categories (for discussion, see Malik, 2001; Heartfield, 2002; Panton, 2012).

Official depictions of the vulnerable subject, as well as radical, progressive challenges to them, resonate well with proclaimed scepticism about the archetypal liberal subject in government behaviour change agendas across different policy terrains. For example, the government's Behavioural Insight

Team predicates its goal of 'changing hearts without changing minds' on images of the disorganised, fallible, slightly (or completely) useless, lazy, indecisive 'Homer Simpson' subject of behavioural economics, who wants to make the 'right' lifestyle choices, is often unhappy that he or she does not or cannot, and therefore needs various paternalistic types of state-sponsored 'nudging', 'shoving' or 'budging' (see Jones *et al.*, 2013; Pykett and Johnson, 2015).

The contemporary appeal of aspirations to bypass rational thought in order to effect desired behaviour change has older roots. For example, between the 1920s and 1950s, advocates of then new ideas from behavioural psychology in schooling, parenting, child guidance and institutional care claimed they did not need to engage with the minds of those unable or unwilling to adapt to such systems (see Thompson, 2006; Myers, 2010; Stewart, 2011). Then, as now, these older accounts reveal no totalizing ideology. For example, advocates of behaviourism in the late nineteenth century and early twentieth century presented it as morally significant, a socially progressive counter both to the determinism of genetics and eugenics that then prevailed and to social advancement bestowed by privilege and nepotism. Conversely, others – notably in state education from the nineteenth century – supported psychological diagnosis, labelling, sifting and categorizing those deemed to be feckless, imbecilic, simple-minded and feeble (and therefore morally deficient) precisely *because* this provided a basis for eugenics and social hygiene (Myers, 2010; Ball, 2013; Allen, 2014).

There are also strong historical parallels to the emphasis by contemporary governments on the need to take account of unconscious, irrational and emotional dimensions of citizens' behaviour. For example, in the early to mid twentieth century, American and British concerns that the authority of the state itself was vulnerable both to democratic resistance and the ineffectiveness of openly rational, educational, politically engaged forms of governance fueled heated interwar debates between sociologists, policy-makers and psychologists about how to sustain democracy whilst containing its potential for resistance and subversion (see Rose, 1999; Furedi, 2013: chapter 8; Nolan, 1998). Here, pessimism about, or openly expressed disdain for, the limited intellectual capacities of an emotionally gullible, volatile, irrational public fueled governments' interest in better, more expansive measures of public attitudes alongside new subliminal propaganda techniques to observe, regulate and manipulate public opinion and morale. Though at their height during the Second World War, insights from this period have continued to update government applications of what Nikolas Rose (1999) calls 'psy-technologies of the self' and thereby to legitimize authority (see also Furedi, 2013: chapter 8; Nolan, 1998).

Competing images of the human subject

Interplays between different images of subjectivity, outlined above, illuminate how different interest groups in different historical periods generate powerful philosophical and political challenges: first to depictions of the rational, reasoning, agentic subject as an effective decision- or choice-maker; and second to the place of subjectivity in a normative framing of democracy based on fundamental universal (or potentially universal) abilities. Understanding some of these interplays also guards against presenting an overly coherent or determinist picture of contemporary behaviour change agendas and associated images of human subjects. Instead, dichotomies between the vulnerable, irrational subject and a liberal or neo-liberal counterpart appear through simultaneous invocations of rationality and coherence alongside the irrational and emotional, the social context and individual mind, the conscious and unconscious, collective and individual, reflexive and unaware, and the resilient and agentic along with the vulnerable. For example, the disciplinary regimes of performance management, league tables, targets, accountability measures and audits that now dominate public and private sector organisations invoke the self-seeking, choosing and resilient neo-liberal subject in tandem with its vulnerable, stressed and anxious counterpart. In educational settings and workplaces, synonymous discourses of 'supporting', 'coaching' and 'developing' the resilience of subjects 'vulnerable to', and therefore anxious about, the operation and outcomes of performative systems have become ubiquitous. In this context, schemes for 'reflective practice', appraisal, mentoring, review and professional/personal development become essential for trying to engineer successful performance in such systems whilst also being a therapeutic respite from them (Ecclestone and Hayes. 2009).

Predictably, competing images of subjectivity also appear amongst advocates of the Behavioural Insight Team's remit. On the one hand, libertarian, paternalistic depictions of 'nudge' techniques denigrate old ideas that governments must 'educate and inform' as a way of engaging with the rational, conscious and agentic self (see Jones *et al.*, 2013; Pykett and Johnson, 2015). On the other hand, some social researchers regard decision-making, engaged and reflexive subjects as essential for the types of deliberative, local choices about the specific focus and subsequent implementation of particular behaviour change projects that will help regenerate community-based democracy (see John *et al.*, 2011). Other contradictions appear in interventions that portray vulnerable subjects as lacking essential psycho-emotional skills and capacities for an increasingly ruthless neo-liberal capitalist system, so that vulnerability is an ever-present yet largely implicit flip side to psychological resilience as invulnerability (Furedi, 2008; Chandler, 2014).

The presentation of dichotomies creates obvious tensions: 'the tendency to inflate the problem of emotional vulnerability and to minimise the ability of the person to cope with distressful episodes runs counter to the therapeutic idealisation of the self-determining individual' (Furedi, 2004: 114). Other tensions are evident. For example, psycho-emotional interventions require 'the individual [to] already be conceived as autonomous and self-reflecting. In other words, the pedagogic process appears to require the pre-existence of that which it is trying to create' (Dahlstedt *et al.*, 2011: 406). The interplay between the rational, capable, reasoning subject and their vulnerable, stressed out, anxious counterpart is ever-present. For example, advocates of positive thinking or 'mental toughness' training programmes appeal to the rational, thinking, reasoning faculties of participants by presenting research evidence which claims that empathy with feelings of being out of control, stressed, disconnected or vulnerable is a springboard for strategies that can reorient these feelings more rationally and thereby more productively (e.g. King, 2014; Cornum, 2012). Similarly, advocates of mindfulness argue that national and regional governments should play an active role in developing engaged, emotionally self-aware, reflexive, psychologically resilient citizens who are able to understand the irrational, emotional, unconscious factors that drive behaviour and then adopt effective strategies to restore a self-aware, ethically authentic self that makes better decisions (Jones *et al.*, 2013; see also Lilley *et al.*, 2014).

Notwithstanding important theoretical and pedagogical differences between interventions cited here, they present a compelling interplay between being an engaged, rational, thinking subject aware of irrational, emotional, unconscious factors that drive behaviour and then able to adopt psycho-emotional strategies to function better, and being able to empathise with its emotional, anxious and vulnerable counterpart. In many ways, then, the ideal human subject envisaged by a therapeutic behaviour change agenda predicated on vulnerability appears to be the rational, individual Cartesian subject who is adaptable, entrepreneurial and self-responsible in the face of neo-liberalism (e.g. Brunila, 2012). Yet, as I argue next, while new types of psy-expert in a growing therapeutic intervention market espouse dichotomies between the 'ideal' and vulnerable subject, the normalisation of the universally vulnerable subject is becoming a necessary foundation for new therapeutic forms of governance.

New therapeutic forms of governance

The rise of a therapeutic market

Multiple, overlapping crises driving behaviour change policy agendas do not merely challenge the legitimacy of capitalism itself. Crises also, of course, challenge the state's authority and competence to address the profound wicked social and individual problems that capitalism creates 'upstream' and which the state must then manage 'downstream' (see Wacquant, 2012). Amidst numerous relentless problematisations of all manner of crisis – from the environment to terrorism, health, social inequality and citizen engagement – highly diffused notions of vulnerability and the potentially universal reach of mental illness as a global 'ticking time bomb' present new challenges to traditional forms of governance. Notably, the state has to legitimise erosions of boundaries between the private and public spheres and a corresponding expansion of specialist and universal interventions in citizens' psycho-emotional states. There is not space here to detail how these challenges and associated government roles have intersected with the rise of a therapeutic culture in legal, welfare and education systems over the past 30 years or so (see Rose, 1999; Nolan, 1998, 2003; Furedi, 2004; Ecclestone and Hayes, 2009). However, a salient point for understanding recent British governments' enthusiasm for a therapeutic behaviour change agenda is the 1997–2010 Labour government's 'third way' approach to social democracy, which promoted an explicitly therapeutic orientation for the British welfare state where 'welfare is not in essence an economic concept, but a psychic one, concerning as it does well-being. ... welfare institutions must be concerned with fostering psychological benefits, as well as economic benefits' (Giddens, 1998: 117).

Reflected in the growing influence of what Jones *et al.* call 'psychocrats' (2013: 82), a shifting, fluid array of therapeutic and psychological expertise has become integral to the state's authority across different behaviour change policy contexts. In his analysis of the intensifying power of psy-experts across popular culture and in government at a time of crisis about the role of the welfare state, Nikolas Rose argued that the decoupling of central state powers from the internal lives of citizens was, in part, rooted in the generosity of the psy-sciences in giving away 'their language, their grammars of conduct and their styles of judgement' (1999: 264). Max Weber's characterisations of different types of authority and shifting relations between them in periods of social change are useful here for understanding the growing intervention market through which a popularised therapeutic behaviour change agenda operates (e.g. Kasler, 1988; Spencer, 1970; see also Furedi, 2013). In particular, his 'ideal types' of traditional, legal and charismatic expert illuminate how this market relies on populist, secular versions of Weber's religious or spiritual charismatic or 'hero' individuals and organisations who eschew traditional science or rule-based foundations for authority whilst also drawing on them for credibility. For example, new psy-expertise in educational settings is reflected in large parallel rises of universal generic programmes administered by teachers, inclusion and engagement workers, learning support and classroom assistants, youth workers and peer mentors, who often have only rudimentary training in their favoured approach, and targeted interventions administered by traditional experts such as educational and clinical psychologists, psychotherapists and trained counsellors. Yet both types of practitioner face growing external competition from publicly funded, short-term projects and commercially marketed programmes administered by charities, campaigning groups, third sector organisations and consultancies (e.g. Rawdin, 2016; Ecclestone and Rawdin, 2016).

These new competitors often present traditional expertise as irrelevant or unhelpful. For example, the Amy Winehouse Foundation – a charity funded by the Big Lottery Fund to run 'resilience training' programmes in schools – proclaims facilitators' own recovery from drug and drink problems as necessary expertise for requiring young people to take part in assessments of their psycho-emotional states and then to participate in interventions. Such activities bought in by schools tend to be sporadic, short-lived and not sustained or followed up (e.g. Rawdin, 2016). Other claims for experience to be the basis for expertise comes from mental health service user campaigns to decentre old types of expertise in favour of peer-based support in mainstream adult and community education programmes that encourage some participants to progress to tutor roles (Lewis, 2014). At a national level, the credibility of the Department for Education's newly created 'mental health czar' for young people is rooted in her own experience of eating disorders and peer mentoring.

In secular versions of the religious or spiritual foundations that Weber thought key to the affective strategies deployed by charismatic experts as the basis for their authority, much promotion of therapeutic behaviour change intervention, regardless of the specific approach, is often through an emotive, enthusiastic, personalised and highly persuasive tone that can be characterised as 'evangelical' (e.g. see artofbrilliance.co.uk; tougherminds.co.uk; thehawn foundation.co.uk; actionforhappiness.org; nlp.org.uk; King, 2014). In the case of the highly popular approach of mindfulness, the secular evangelism that promotes it invokes its Buddhist spiritual roots (e.g Lilley *et al.*, 2014).

Importantly, both everyday life and policy-based charismatic experts gain legitimacy from their more famous counterparts. Examples include the actor Goldie Hawn and comedian Ruby Wax who proclaim the necessity of school-based, universal mindfulness programmes, and United States Army general Rita Cornum who promotes a lay version of her army's 'resilience fitness training' for education and youth work services (hawnfoundation.org; Wax, 2013; Cornum, 2012). Closer to Weber's ideal type, in their campaigns for public and political influence, they proselytise the personal benefits they have gained from a particular approach and claim that these should be offered universally rather than as targeted responses. At the level of policy creation, charismatic experts in left-liberal think tanks such as DEMOS, The Young Foundation and the Royal Society of Arts have driven the growing relationship between government advisory roles and advocacy of various types of psychologically oriented state intervention (see Frawley, 2014; Jones *et al.*, 2013). Outside of key individuals in particular policy organisations, high-profile charismatic experts have advised successive British governments; including Martin Seligman, leading authority on positive psychology and former president of the influential American Psychological Association (which created the *Diagnostic and Statistical Manual of Mental Disorders*), and Richard Layard, Professor of Economics, founder of the high-profile Action for Happiness campaign and 'happiness czar' for Labour governments (Seligman, 2011; actionforhappiness.org). Perhaps the most famous archetypal charismatic hero expert is the Dalai Lama who endorses the Action for Happiness campaign for the use of mindfulness in public policy (Easton, 2015).

Destabilising therapeutic governance

From a Weberian perspective, these new types of psy-authority both encourage and are reinforced by popularised forms of practical or technical rationality that present universal vulnerability and the threat of mental illness as an 'ethical standard … a specific type of value-rational belief which … imposes a normative element on human action that claims the quality of the "morally good"' (Karlberg, 1980: 1165). As noted above, influential claims to authority and expertise in the therapeutic intervention market gain credence from long-running advocacy in critical psychology and user or survivor campaigns for peer-based, grass-roots approaches. In this context, calls for vulnerability to be the 'moral standard' of an inclusive, progressive state that recognises and addresses the oppressions of neo-liberal capitalism are predicated on rejecting the 'folly' of the rational subject as a basis for governance (Fineman, 2015). The privileging of vulnerability intertwines both with radical rejections of traditional expertise and a popular sensibility that anxiety and vulnerability are part of an inexorable continuum of a mental illness epidemic, experienced or feared as both personal and social crisis.

The legitimacy of a state-sponsored therapeutic market also comes from its location in an established and growing patchwork of state-funded, privatised education provision (Ball and Junnermann, 2012). In her exploration of projects that encourage citizens to initiate certain politically motivated 'pedagogies', Janet Newman (2012) argues that ethical discourses of social justice, human rights, progressivism, empowerment, self-help and peer support enable radical/ progressive/critical standpoints to play a crucial role in state-sponsored governance. Seen in this light, the decentring of psy-expertise and the politicisation of the universally vulnerable, anxious subject are a compelling contrast to old types of psy-experts working through regulated, hierarchical forms of governance. Rather than seeking to remedy or apologise for the negative effects of liberal universalism or the oppressions of neo-liberalism, charismatic non-specialists offer culturally familiar, seemingly non-judgemental therapeutic strategies to deal with new problems in a complex, uncertain world. Specifically, the political transmogrification of vulnerability from a stigma or deficit attached to a degraded liberal and neo-liberal subject enables a reframing of state bureaucracy into populised and marketised therapeutic forms.

Yet, despite advantages for the state in opening up old forms of psy-expertise, myriad claims to authority and legitimacy also create risks. Here, the characteristics that make therapeutic governance credible and appealing – namely, being accessible, short-term, responsive and culturally resonant – also make it unstable, shifting and largely unaccountable. In turn, the evidence base becomes fragmented, short-term and small-scale and, through its promotion by advocates or implementers of particular interventions, prone to confirmation bias. I turn next to some of the dangers that arise from these characteristics of therapeutic forms of governance.

New dangers

Governing through neurosis

Engin Isin's Foucauldian account of the ways in which governments create and respond to neurosis helps to illuminate dangers generated by the forms of therapeutic governance I have outlined above. Responding critically to what he sees as shortcomings in sociological accounts of 'risk societies', Isin argues that the neurotic citizen is incited to make social and cultural investments to eliminate various dangers by calibrating its conduct on the basis of its anxieties and insecurities. Invited to consider itself as part of a neurological species and then to understand itself as an affect structure, the subject at the centre of governing practices is not understood as competent, where, with relative success, it can evaluate alternatives by using its rational capacities for managing truth claims to avoid or eliminate risks. Instead, an inherently anxious, stressed-out, individualized and increasingly insecure subject is asked to manage its neurosis and associated anxieties through affects. According to Isin, this mobilizes a kind of power through agency to seek freedom from anxiety. Using examples of public and government depictions of crises in economic management, the body and the environment, Isin argues that 'relentlessly repetitive problematisa-tions' target anxiety, not just amongst citizens as individuals but also the various groups and individuals who govern them (2004: 227). For example, studies of eating disorders illuminate how young women are constituted as objects of government, not through the bodies and minds of young women themselves but by addressing those who govern them – such as teachers, social and youth workers, medical professionals, and carers/parents – as anxious subjects whose actions will affect those for whom they are responsible.

An infinite constituency of anxious subjects?

Relating Isin's argument to discussion in this chapter, the cultural and political intertwining of relentless problematisations of mental health, vulnerability and resilience produce powerful moral imperatives – Weber's ethical standard or value rationality – for intervention. Yet unlike many other types of problem-atisations, their underlying moral or ethical standard targets an infinite con-stituency of anxious subjects. These include traditional psy-experts and a growing array of professionals, paraprofessionals and non-specialists and their political sponsors. A therapeutic culture and its translations into a therapeutic behaviour change agenda also lends anxiety about mental health a particular personal potency that draws in structural and social risks as sources of psycho-emotional distress alongside repressed or hidden risks – 'emotional baggage' – that lurk in the legacies of our emotional past and present unless we understand and deal with them. This fear encompasses us as individuals whilst also extending out-wards to psychologically dysfunctional families deemed to wreak social havoc unless the state intervenes (see Allen, 2011; Field, 2010; Layard in Sharples, 2007). In turn, the de-stigmatised language of mental health 'issues', in place of mental illness, fuses with highly generalized understandings of vulnerability to fuel apocalyptic fears that early years settings and schools must address these in order to prevent serious future social problems (see Morgan, 2015a, 2015b).

Returning to Isin's thesis, a cycle emerges: striving for a perfect mind or body creates neurotic subjects (specifically those in charge of them), neurotic problematisation in search of solutions through neuro-psychologisation (such as medicalization), followed by increasing reliance on neurogenetic science and its role in expanding diagnoses of emotional, psychological and behavioural disorders. Yet, as I have argued, the populist market of therapeutic governance that emerges from this cycle reaches far beyond specialist claims from neuro-psychologisation, behavioural science and therapy and also beyond the vested interests of a market of interventions and psychotropic drugs. Similarly, critical, radical alternatives to those vested interests and traditional expertise extend a long way beyond grass-roots demands for peer-based solutions to particular mental health problems to a privileged place as universal or generic interventions in a therapeutic market. Increasingly, these approaches are rarely either voluntary or offered in response to requests for intervention. Instead, compelling appeals to the moral responsibilities of professionals, paraprofessionals and parents, together with more generalized anxiety about vulnerability and mental health, combine with the expansion of psy-expertise into popularized lay forms. This creates two significant slippages from specialist interventions in response to a perceived or diagnosed need: first, to the idea that interventions can benefit everyone even if a need is not expressed and then, encouraged by the normalisation of vulnerability, to a perception that psycho-emotional governance is necessary for functioning in a widening range of life situations. Returning to Rose's (1999) claim that the state decouples or disperses its responsibility for the inner lives of citizens to citizens themselves, a therapeutic market seems not to do either of these; rather, it expands the state's role in both specialist and everyday general psychological governance through a widening constituency of anxious subjects.

Actively neurotic citizens

It is important to recognize that anxious subjects conscious of their own psycho-emotional vulnerability or feeling potentially or actually mentally unwell or ill are not a mere social construction. It is not therefore sufficient to argue that everyday affective states are simply being redefined as mental health problems. Rather, neurosis and anxiety about vulnerability and stress, as well as a general sense of fragile psycho-emotional states, are real, embodied and deeply felt. For example, Will Davies (2015) argues that burgeoning workplace stress and low-level mental health problems reflect a deeper cultural psychological malaise that governments and employers respond to. At the same time, however, the neurotic or vulnerable subject is not a passively pathologised target but an active agent (Isin, 2004). According to Isin, one of the most dangerous movements in our times is the articulation of neurotic claims by neurotic citizens where, striving to eliminate anxiety, they evolve a highly sensitized sense of entitlement that it is a matter of justice not to suffer from anxiety. Here, people may extend an empathic sense of justice to others but also make others responsible for their anxiety (Isin, 2004).

In trying to explore complex, sometimes contradictory, iterations between mental health and vulnerability as social construct, agency and embodied experience, I am mindful that such explorations are sensitive and contested. For example, while new forms of therapeutic governance might invoke empathy with others' vulnerability, they also create ever-present fears that others threaten our psycho-emotional equilibrium, competing claims to a vulnerable identity and then therapeutic resources that support or ameliorate that identity (see also McLaughlin, 2011). In this context, the actively neurotic citizen sensitized to vulnerability, together with the embodiment of circular cultural anxieties and the often apocalyptic problematisations that fuel these are also contagious in priming individualized feelings of vulnerability, poor resilience and impending mental illness.

Compulsory therapeutic governance

Iterations between cultural neurosis and anxiety and new forms of therapeutic governance also risk the imposition of interventions on subjects problematized as vulnerable and the normalization of expectations that intervention is a necessary response. This should not suggest that some interventions are not useful or effective for some participants, even if they do not always request them. Yet as I have already argued, examples highlighted above extend a long way past voluntary participants, responses to expressed need or grass-roots demand; instead, participation is increasingly compulsory or assumed. Indeed, in some instances, non-participation, lack of enthusiasm or failure to express oneself in the 'right' way can lead to speculations about repressed vulnerability and potential mental health 'issues' (e.g. Rawdin, 2016).

These characteristics of therapeutic governance open opportunities for more repressive approaches. For example, analyses of the government's Prevent strategy show the dangers that arise from its statutory demand for school teachers and youth workers, and, more recently, university lecturers and support staff, to identify young people deemed 'vulnerable to radicalisation' (Coppock and McGovern, 2014; Richards, 2011; Durodie, 2016). Followed by detailed lists of 'radicalisation narratives' and highly dubious, generalised psychological 'indicators' of radicalization, including 'relevant mental health issues', Prevent portrays human subjects incapable of rational, reasoned consideration of political views or moral responsibility for these. The desirability or otherwise of those views is not the point here; rather, Prevent's sweeping account of what counts as 'radical' together with its explicit couching of radicalization in terms of psycho-emotional vulnerability and generalised reference to 'mental health issues' draw in non-psy-experts to these discourses and activities. The wider eschewing of the rational, reasoning, autonomous subject, discussed in this chapter, legitimizes Prevent as a particularly intrusive example of statutory psychological governance.

Conclusions

There has not been space in this chapter to do justice to complex, often intractable, ontological and epistemological explorations of and disagreements about the foundations of subjectivity reflected in the different histories, historical contingencies, political motivations and research genres of these debates.

There is therefore a related danger that my arguments here are seen, wrongly, as a classed, ableist, gendered and raced privileging of those subjects, or as a simplifying of the relationship between rational, cognitive, affective and emotional dimensions to subjectivity (e.g. Ecclestone and Goodley, 2014; Leathwood and Hey 2009). Nor has it been possible here to account fully for the treatment in these debates of the relationship between the rational and affective dimensions of subjectivity. Similarly, my account of competing claims to psy-authority should not imply a privileging of traditional expertise or suggest that peer-based and lay approaches do not offer a legitimate basis for useful and effective therapeutic approaches. Rather, interventions construct and respond to subjects, and subjects reciprocate, in complex, contradictory and shifting ways. Effects cannot therefore be characterised through binaries between, for example, empowering or repressive, normalising or emancipatory (see Ecclestone and Brunila, 2015).

To some extent, new forms of therapeutic governance discussed in this chapter indicate that the archetypal, idealized liberal citizen and its neurotic version are not exclusive, independent subjects, but produce each other (Isin, 2004). Yet the examples I have presented also suggest that espoused dichotomies between those subjects belie a significant shift in contemporary governance. By responding to citizens' seemingly intractable psycho-emotional problems through *ad hoc*, unstable and largely unaccountable forms of therapeutic governance, the state and its diverse psy-experts seem not to see their target as the archetypal subject of liberal and neo-liberal governance. This suggests that therapeutic governance is not an apology for the failings of liberal universalism and the inequalities of neo-liberalism; rather, it embraces the anxious subject as a natural, necessary, everyday form of governmentality. In this context, radical alternatives to traditional psy-expertise and calls for vulnerability to be a progressive political project are integral to the normalizing of expectations that constant psychological adaptation is necessary for dealing with a complex, changing, troubling world. I have aimed to show that dangers arise when calls from progressive, radical, critical standpoints for vulnerability to transgress or challenge liberal/modernist hierarchies and their outmoded, oppressive privileging of the rational subject are appropriated easily by the state, often using very similar discursive social justice inflections.

In place of any totalizing or determinist ideology or 'real' target of governance by a knowing state, the powerful cultural and political resonance of vulnerability suggests a state increasingly uncertain about its ability to govern through ideas about the liberal and neo-liberal subject that it sees as inappropriate, mythical or outdated (e.g. Wacquant, 2012; Chandler, 2014). Seen in this light, the evangelic, compelling proclamations of a therapeutic intervention market appear to offer the state a response to crisis. Yet this market also exposes governments to constant, competing and sometimes, I would argue, highly dubious claims to have legitimate psy-expertise. Circular depictions of citizens as vulnerable (or of citizens who present themselves as such), instability and uncertainty create new dangers in the form of questionable expertise and evidence and associated dangers that psychological governance is compulsory, unthinking and potentially repressive.

Acknowledgements

I am very grateful to colleagues who have taken the time to offer important insights and challenges in response to an earlier draft of this chapter: Ansgar Allen (Sheffield), Kate Brown (York), Kristiina Brunila (Helsinki), David Chandler (Westminster), Ashley Frawley (Swansea), Dan Goodley (Sheffield), Ken McLaughlin (Manchester Metropolitan University) and Lisa Proctor (Manchester Metropolitan University).

References

Allen, A. (2014) *Benign Violence: Education in and Beyond an Age of Reason*. Palgrave Macmillan, London.

Allen, G. (2011) Early Intervention: The Next Steps. An Independent Report to Her Majesty's Government. Department for Education, London.

Bache, I., Reardon, L. and Anand, P. (2015) 'Wellbeing as a wicked problem: navigating the arguments for the role of government'. *Journal of Happiness Studies*, 17(3): 893–912.

Ball, S. J. (2013) *Foucault, Power and Education*. Routledge, London.

Ball, S. J. and Junnermann, C. (2012) *Networks, New Governance and Education*. Routledge, London.

Brown, K. (2015) *Vulnerability and Young People: Care and Control in Social Policy*. Policy Press, Bristol.

Brunila, K. (2012) 'A diminished self: entrepreneurial and therapeutic ethos operating with a common aim'. *European Educational Research Journal*, 11(4): 477–486.

Chandler, D. (2014) *Resilience: The Governance of Complexity*. Routledge, Abingdon.

Coppock, V. and McGovern, M. (2014) '"Dangerous minds"? Deconstructing counterterrorism discourse, radicalisation and the "psychological vulnerability" of Muslim children and young people in Britain'. *Children and Society*, 28(3): 242–256.

Cornum, R. (2012) 'Can we teach resilience?' Keynote presentation to Young Foundation/Macquarie Group seminar, Teaching Resilience in Schools, Ropemaker, London, 7 February.

Dahlstedt, M., Fejes, A. and Schönning, E. (2011) 'The will to (de)liberate: shaping governable citizens through cognitive behavioural programmes in school'. *Journal of Education Policy*, 26(3): 399–414.

Davies, W. (2015) *The Happiness Industry: How Government and Big Business Sold Us Happiness and Well-being*. Verso, London.

Department for Education and WAVE Trust (2013) From Conception to Age 2: The Age of Opportunity. Department for Education, London.

Durodie, B. (2016) 'Securitising education to prevent terrorism or losing direction?' *British Journal of Educational Studies*, 64(1): 21–35.

Easton, M. (2015) 'Evening classes that promise to make you happy' [online]. *BBC News*, 21 September. Available at: www.bbc.co.uk/news/uk-34292274

Ecclestone, K. (Ed.) (2013) *Emotional Well-being in Policy and Practice: Interdisciplinary Perspectives*. Routledge, London.

Ecclestone, K. and Hayes, D. (2009) *The Dangerous Rise of Therapeutic Education*. Routledge, London.

Ecclestone, K. and Goodley, D. (2014) 'Political and educational springboard or straitjacket? Theorising post/human subjects in an age of vulnerability'. *Discourse: Studies in the Cultural Politics of Education* [published online 17 June]. DOI: 10.1080/01596306.2014.927112.

Ecclestone, K. and Lewis, L. (2014) 'Interventions for emotional well-being in educational policy and practice: challenging discourses of "risk" and "vulnerability"'. *Journal of Education Policy*, 29(2): 195–216.

Ecclestone, K. and Brunila, K. (2015) 'Governing emotionally vulnerable subjects and "therapisation" of social justice'. *Pedagogy, Culture and Society*, 23(4): 485–506.

Ecclestone, K. and Rawdin, C. (2016) 'Reinforcing the "diminished" subject? The implications of the "vulnerability zeitgeist" for well-being in educational settings'. *Cambridge Journal of Education* [published online 14 January]. DOI: 10.1080/0305764X.2015.1120707.

Field, F. (2010) The Foundation Years: Preventing Poor Children Becoming Poor Adults. Independent Review on Poverty and Life Chances. Department for Work and Pensions: London.

Fineman, M. (2008) 'The vulnerable subject and the responsive state'. *Yale Journal of Law and Feminism*, 20(1): 1–22.

Fineman, M. (2015) 'Social justice and the vulnerable state'. Inaugural lecture at the Centre for Law and Social Justice, University of Leeds, 16 September.

Frawley, A. (2014) *The Semiotics of Happiness: The Rhetorical Beginnings of a Social Problem*. Bloomsbury, London.

Furedi, F. (2004) *Therapy Culture: Cultivating Vulnerability in an Uncertain Age*. Routledge, London.

Furedi, F. (2008) 'Fear and security: a vulnerability-led policy response'. *Social Policy and Administration*, 42(6): 645–661.

Furedi, F. (2013) *Authority: A Sociological History*. Cambridge University Press, Cambridge.

Giddens, A. (1998) *The Third Way: The Renewal of Social Democracy*. Polity, Oxford.

Gillies, V. (2015) 'Troubled families: lessons from historical comparative analysis'. Paper presented to the Social Policy Association annual conference, Social Policy in the Spotlight: Change, Continuity and Challenge, University of Ulster, Belfast, 6–8 July.

Harwood, V. and Allan, J. (2014) *Psychopathology at School: Theorizing Education and Mental Disorder*. Routledge, London.

Heartfield, J. (2002) *The Death of the Subject Explained*. Sheffield Hallam University, Sheffield.

Isin, E. (2004) 'The neurotic citizen'. *Citizenship Studies*, 8(3): 217–235.

John, P., Cotterill, S., Hahua, L., Richardson, L., Moseley, A., Smith, G., Stoker, G. and Wales, C. (2011) *Nudge, Nudge, Think, Think: Using Experiments to Change Citizens' Behaviours*. Bloomsbury, London.

Jones, R., Pykett, J. and Whitehead, M. (2013). *Changing Behaviours. On the Rise of the Psychological State*. Elgar Publishing, Cheltenham.

Jopling, M. and Vincent, S. (2015) Vulnerable Children: Needs and Provision in the Primary Phase. Unpublished report for the Cambridge Primary Review Trust. Northumbria University, Newcastle.

Karlberg, S. (1980) 'Max Weber's types of rationality: cornerstones for the analysis of rationalization processes in history'. *American Journal of Sociology*, 85(5): 1145–1179.

Käsler, D. (1988) *Max Weber: An Introduction to His Life and Work*. University of Chicago Press, Chicago.

King, V. (2014) Building resilience – practical interventions to help people survive and thrive in today's world. Presentation to ESRC Seminar Series, Psychological Resilience: Governing the Brain, Mind and Behaviour, University of Birmingham, 23 June.

Leathwood, C. and Hey, V. (2009) 'Gendered discourses and emotional sub-texts: theorizing emotion in UK higher education'. *Teaching in Higher Education*, 14(4): 429–440.

Lewis, L. (2014) 'Responding to the mental health and well-being agenda in adult and community learning'. *Research in Post-Compulsory Education*, 19(4): 357–377.

Lilley, R., Whitehead, M., Howell, R., Jones, R. and Pykett, J. (2014) Mindfulness, Behaviour Change and Engagement in Public Policy: An Evaluation. University of Aberystwyth, Aberystwyth.

McLaughlin, K. (2011) *Surviving Identity: Vulnerability and the Psychology of Recognition*. Routledge, London.

McLeod, J. (2012) 'Vulnerability and the neo-liberal youth citizen: a view from Australia'. *Comparative Education*, 48(1): 11–26.

Malik, K. (2001) *Man, Beast or Zombie?* Weidenfield, London.

Miller, P. and Rose, N. (2008) *Governing the Present: Administering Economic, Social and Personal Life*. Polity Press, Cambridge.

Mills, C. (2014) *Decolonising Global Mental Health: The Psychiatrisation of the Majority World*. Routledge, London.

Morgan, N. (2015a) Speech to Early Intervention Foundation Conference, 12 February. Available at: https://www.gov.uk/government/speeches/nicky-morgan-speaks-at-early-intervention-foundation-conference (accessed 4 July 2015).

Morgan, N. (2015b) 'Make happiness a priority in schools. Report on visit by the Secretary of State for Education to Upton Cross School'. *The Times*, 4 July, p. 7.

Myers, K. (2010) 'Contesting certification: mental deficiencies, families and the state'. *Paedagogica Historica: International Journal of the History of Education*, 47(6): 749–766.

Newman, J. (2012) 'Toward a pedagogical state? Summoning the "empowered" citizen', in J. Pykett (Ed.). *Re-educating Citizens. Governing through Pedagogy*. Routledge, London, pp. 95–108.

Nolan, J. L. (1998) *The Therapeutic State: Justifying Government at Century's End*. New York University Press, New York. Nolan, J. L. (2003) *Reinventing Justice: The American Drug Court Movement*. University of Princeton Press, Princeton, NJ.

O'Donnell, G., Denton, A., Halpern, D., Durand, M. and Layard, R. (2014) Well-being and Policy. Report for the Legatum Institute. London School of Economics, London.

Office for Standards in Education (2012) Good Practice in Creating an Inclusive School. Department for Education, London.

Panton, J. (2012) The Politics of Subjectivity. PhD thesis, University of Oxford, Oxford.

Paterson, C., Tyler, C. and Lexmond, J. (2014) Character and Resilience Manifesto. All-Party Parliamentary Group on Social Mobility, CentreForum and Character Counts, London.

Paur, J. (Ed.) (2012) 'Precarity talk: a virtual roundtable with Lauren Berlant, Judith Butler, Borjana Cvejić, Isabell Lorey, Jasbir Paur, Ana Vujanović'. *The Drama Review*, 56(4): 163–177.

Pykett, J. and Johnson, S. (2015) Silver Bullets Need a Careful Aim: Dilemmas in Applying Behavioural Insights. University of Birmingham, Birmingham.

Rawdin, C. (2016) Social and Emotional Learning Interventions and the Reshaping of Teachers' Subjectivity. PhD thesis, University of Birmingham, Birmingham.

Richards, A. (2011) 'The problem with "radicalisation": the remit of "Prevent" and the need to refocus on terrorism in the UK'. *International Affairs*, 87(1): 143–152.

Richardson, L. (2011) 'Cross-fertilisation of governance and governmentality in practical policy making on behaviour change'. *Policy and Politics*, 39(4): 433–446.

Rose, N. (1999) *Governing the Soul. The Shaping of the Private Self*, second edition. Free Association Books, London.

Seligman, M. E. (2011) *Flourish: A New Understanding Of Happiness And Well-being – And How to Achieve Them*. Simon Shuster Books, New York.

Sharples, J. (2007) Well-Being in the Classroom. Report on the All-Party Parliamentary Committee on Scientific Research in Learning and Education. University of Oxford, Oxford.

Social Exclusion Task Force (2007) Families at Risk: Background on Families with Multiple Disadvantages. Cabinet Office, London.

Social Exclusion Unit (1999) Bridging the Gap: New Opportunities for Young People Not in Education, Employment or Training. Cm 4405. Social Exclusion Unit, London.

Spencer, M. E. (1970) 'Weber on legitimate norms and authority'. *British Journal of Sociology*, 21(2): 123–134.

Stewart, J. (2011) 'The dangerous age of childhood: child guidance and the "normal" child in Great Britain, 1920–1950'. *Paedagogica Historica: International Journal of the History of Education*, 47(6): 785–803.

Thompson, M. (2006) *Psychological Subjects: Identity, Culture and Health in Twentieth Century Britain*. Oxford University Press, Oxford.

Valentine, G. and Skelton, T. (2003) 'Living on the edge: the marginalization and "resistance" of D/deaf youth'. *Environment and Planning*, 35(2): 301–321.

Wacquant, L. (2012) 'Desperately seeking neo-liberalism: a sociological catch'. Keynote presentation to The Australian Sociological Association annual conference, Emerging and Enduring Inequalities, University of Queensland, 11 November.

Wax, R. (2013) *Sane New World: Taming the Mind*. Hodder and Stoughton, London.

Wright, K. (2011) *The Rise of the Therapeutic Society: Psychological Knowledge and the Contradictions of Cultural Change*. Academia Publishing, New York.

5 Psychology as practical biopolitics

The Midlands Psychology Group: John Cromby,
Bob Diamond, Paul Kelly, Paul Moloney,
Penny Priest and Jan Soffe-Caswell

Introduction

Psychology today is a technique of government. Some analysts (e.g. Rose, 1985) propose that this has always been the case, that since its inception, psychology has been implicated in practices of inscription, measurement and calculation designed to render populations more manageable. Rose argues that the mere existence of psychology as a separate discipline already exemplifies a particular, individualising tool of governance. Whilst the generality of this claim might be questioned (perhaps, to some extent, psychology has also become more than this), it is nevertheless clear that, in recent years, psychology has increasingly been taken up quite overtly by governments, particularly within their efforts to manage and promote good health and healthy lifestyles. Behavioural change, conceived in largely individualised psychological terms (as opposed to being consequential upon societal, structural influences) is the focus of various contemporary initiatives to foster health and promote wellbeing (Crawshaw, 2013). Simultaneously, psychological evidence and practices are central to the work of the UK Behavioural Insights Team (BIT) or 'Nudge Unit'. Initially a Cabinet Office initiative, the BIT is now a social enterprise company whose aims include 'encouraging' people to 'make better choices for themselves and society' (Behavioural Insights Team, 2015) by drawing upon specialist knowledge from disciplines including psychology. Health and wellbeing are frequently linked to happiness with claims sometimes made that happiness itself fosters good health (although evidence for this causal link is shaky; see Pressman and Cohen, 2005). Psychological notions such as self-esteem, coping and resilience are also centrally implicated within the so-called 'assets-based' approach to public health (Friedli, 2013; Mindfulness All-Party Parliamentary Group, 2015). Hence, in frequent association with positive psychological concepts, governmental (and indeed some extra-governmental – see New Economics Foundation, 2004) initiatives work with quite explicitly psychologised notions of happiness and wellbeing.

In what follows, then, we consider the nexus of psychology and health with relation to governance. We do so by focusing closely upon the methodological practices used to generate psychological research evidence, taking a specific paper and its findings as our example. We frame our analysis with respect to Foucault's notion of biopolitics because this concept makes explicit the intimate links between health and governance in the current neoliberal era. Comparisons and contrasts might nevertheless usefully be drawn between our analysis and others in this volume, such as the analyses of happiness by Binkley and Davies that distinguish between the disciplinary governance Foucault posited and the more fluid and dispersed control societies emphasised by Deleuze (1992). We begin by outlining the concept of biopolitics; we then introduce the study being analysed, outline some general points with regard to its methods and present a more specific analysis of the paper in question. We conclude by describing some possible implications of our analysis for research and for practice.

Biopolitics and contemporary governance

Biopolitics, at least in relation to Foucault's (2008) work (the term also has more general meanings), is a complex and occasionally contested concept. In Foucault's writings, another concept, governmentality, is frequently linked with biopolitics. Governmentality refers both to the ways that governments strive to produce subjects best fitted to their own values and intentions, and to the various organised practices and techniques by which subjects are governed. For Foucault, biopolitics is also connected to the development and operation of the practices of neoliberalism. In close relation to these other concepts, then, biopolitics describes *a style of governance that engages with life itself,* where life is understood quite broadly as arrayed along a spectrum between the vicissitudes and vulnerabilities of the individual body at one extreme and the abstract or aggregate (statistical – i.e. 'of the state') qualities of the entire population at the other.

Foucault locates the emergence of this style of governance in nineteenth-century Europe, where at the individual level it is associated with techniques and practices of discipline, and at the population level with techniques of observation, calculation, normalisation, rationalisation and legitimation. To some extent, as Lemke (2016) observes, this may overstate the historical and cultural specificity of biopolitics; other scholars claim that its origins can be both traced back to Greek antiquity and discerned far beyond the Western hemisphere. Lemke notes that the specificities of Foucault's concept of bio-politics have also been questioned with regard to the forms of (disciplinary) power it implicates, the political rationalities it presumes, and the concepts of the body upon which it depends. Notwithstanding these questions, Foucault's concept is widely seen as having considerable analytic purchase.

In immediately practical terms, for Foucault, the emergence of biopolitics signalled a shift in emphasis away from sovereign power, characterised pre-dominantly as the right to take life or let live, toward a greater emphasis upon 'biopower': the right to make live and let die. Crudely put, the rise of biopo-litics (and the biopower it enacts) represented a move away from an earlier 'off with their heads' mode of government, where relatively savage and spectacular punishment was the predominant means of exercising power, toward a more nuanced mode of exerting power whereby individuals and populations are worked upon more subtly, actively, and continuously. Thus, rather than understanding power almost exclusively as 'interdiction and repression', from a biopolitical perspective, power is also understood in *productive* terms 'dedicated to inciting, reinforcing, monitoring and optimizing the forces under its control' (Lemke, 2016: 57).

It is perhaps through enacting and regulating the right both to make live and to let die that biopolitics is most intimately linked to governmentality. In contemporary neoliberal societies, ideological, ethical and moral imperatives get intertwined with discourses taken as ways of speaking truth about life itself: discourses associated with disciplines including genomics, epidemiology, neuroscience or – the focus of this chapter – psychology (specifically, its measurement practices). Each of these discourses foregrounds different aspects of life, locating its dynamism variously within the DNA, the population, the brain or the thinking of the individual. In so doing, they each yield apparent truths: these truths then function as raw materials and sources of normative expectation by and through which individual subjectivities might be constituted and in accord with which the rights to make live and let die are legitimated, organised and enacted.

Under neoliberalism, individuals are encouraged to position themselves as choosing rationally in relation to these apparent truths, making sound use of the best information and advice in order to conduct their lives responsibly. These tropes of rational choice and responsibility sit alongside related neoliberal emphases including entrepreneurship, competitiveness and individualism. Their confluence frequently acquires heightened significance because neoliberalism's rolling back of welfare and health provision means that the adverse material consequences of appearing to choose badly, or of seeming to lack personal responsibility, entrepreneurial flair or competitiveness, become more severe. Neoliberal emphases upon individuality, personal responsibility, competition and choice precisely but inversely reflect the withdrawal and denigration of societal provision, collective support and governmental responsibility with which they are temporally concurrent.

Empirically, psychology is often prominent within the knowledges, practices and techniques deployed by neoliberal governments in their efforts to forge subjectivities consonant with their own interests. In the UK, examples include the Blair government's adoption of Layard and colleagues' proposals (in Centre for Economic Performance's Mental Health Policy Group [2006]) to extend the provision of psychological therapy under the IAPT (Improving Access to Psychological Therapies) programme. The ostensibly humane and compassionate dimensions of this policy are undercut not only by the questionable efficacy of the therapies promoted (Moloney, 2013) but also by the linking of therapeutic engagement with welfare entitlement. This connection was there from the outset: Layard and colleagues' initial proposals (Centre for

Economic Performance's Mental Health Policy Group, 2006) directly equated the cost of training and employing sufficient therapists to 'cure' depression and anxiety amongst benefit claimants with the cost of maintaining those same people on benefits. But as IAPT has progressed, the linking has become more explicit, with benefit entitlements for some now contingent upon weekly attendance at therapy sessions. Similarly, psychology was central to an initiative under the 2010 UK Coalition government (and devised by the BIT) that required benefit claimants to undergo psychological testing in order to identify their 'signature strengths' and so – by some process never elucidated – improve their chances of finding work. In fact, as Cromby and Willis (2014) demonstrated, both the content of the test and the procedure it entailed were able to function as techniques capable of reinforcing specific aspects of neoliberal subjectivities.

It is important to recognise that subjectivities have the continuous potential to escape or exceed the normative and ideologically inflected knowledges and practices that regulate them. In the rich flux of everyday life, neoliberal precepts and practices frequently encounter social and material contradictions that stymie them, omissions that render them superfluous, or situations and events that challenge their claims to truth and acceptability. Experience is therefore constantly being opened up by necessary excesses that transcend, subvert or deny neoliberal and other ideologically inflected edicts (Stephenson and Papadopoulos, 2007). If neoliberalism were simply, uncontrovertibly 'correct', determined attempts to instil its precepts within subjectivity would be unnecessary; hence, the very existence of initiatives such as IAPT and bodies such as the BIT demonstrates the *limits* of neoliberalism – thereby revealing the ongoing ideological struggles associated with it.

Understood in relation to governmentality, then, psychology is both a terrain of contestation for neoliberal ideologies and, at the same time, a set of resources (practices, knowledges, techniques) whereby subjects are variously incited to understand themselves (and their relationships, activities, aspirations, ideals and potentials) in predominantly neoliberal terms. With particular reference to biopolitics, psychology is a source of knowledge and practice with regard to policies and initiatives designed to make live: to impel people to live in particular ways, to endorse certain views of the good or responsible life, to promote and normalise ways of achieving this good life. Implicitly, at least, this means that psychology must also function as a resource that prescribes the less good life – those ways of living characterised by poor choice and failed responsibility. In complementary fashion, then, and with respect to those who either will not or cannot enact sufficient responsibility, psychology furnishes explanations (that can function simultaneously as justifications) that might allow these irresponsible people to be individualised, problematised, and thus – if not actually left to die – at least discounted, as authors of their own misfortunes, from any beneficent responsibility of government.

To illustrate how psychology can function in this manner as a form of practical biopolitics, in this chapter, we examine an individual piece of psychological research: a study by Packard *et al.* (2012) which claimed to show that personality traits influence non-compliance with regimes of healthy exercise and diet amongst people living in deprived areas of a Scottish city. This study was chosen because it exemplifies especially clearly both the individualising tendencies of (most) psychological research and the ways in which psychometric methods function to reproduce and 'real-ise' these tendencies through the provision of apparently value-free 'scientific' evidence. Close engagement with the detail of research is necessary to demonstrate this, so a sustained focus on a single study is required. This absolutely does not mean, however, that this study is necessarily any more methodologically or conceptually questionable than other psychological research in this field. Our detailed examination of this study is intended to clearly demonstrate the functioning of the sets of interlocking presuppositions, conceptual and methodological, that underpin this genre of research; it is not an indictment of this particular study.

Like other research of this kind, Packard *et al.*'s depended upon psychological assessments of constructs including personality, self-esteem and wellbeing, for the most part using well-established measures. There are many circumstances in which psychological measurement and assessment is both valuable and warranted (e.g. to identify cognitive deficits following stroke or brain injury). But in other circumstances, its utility, efficacy and value are far less clear, and we suggest that Packard *et al.*'s study is more accurately seen as a biopolitical instantiation of neoliberal ideology than as an objective scientific study. To demonstrate this, we first present a general critical discussion of the practices of psychological measurement upon which Packard *et al.*'s analysis depends; we then present a more focused discussion of the specific procedures and measures they describe.

Measuring the psyche

Today, practices of psychological measurement, *psychometrics*, are so ubiquitous and mundane that few give pause to consider their strangeness and peculiarity. Market research surveys (in the street, in shops, online), assessments in self-help books, 'know yourself' questionnaires in newspapers and magazines, and government initiatives such as the national wellbeing survey have all functioned to normalise psychological measurement. Their proliferation means that, from an early age, citizens of contemporary democracies are implicitly schooled to accept the legitimacy of psychological techniques that are, nevertheless, of questionable validity; that is, to the extent that they do measure something, it is not always clear that they measure what they claim. Likewise, despite sometimes being prioritised at the expense of validity, their reliability (i.e. their ability to generate consistent scores in similar circumstances) is also frequently less than ideal. Illustrating this, we now briefly summarise six general concerns associated with psychometrics, all of which are relevant to Packard *et al.*'s study.

First, a great many psychometric measures, including those used by Packard *et al.* (2012), depend wholly or largely upon self-reports. With some small and non-essential differences, such measures face respondents with questions or statements crafted by a psychologist. Respondents are requested to answer the questions or to indicate their level of agreement with the statements. Responses are recorded using grids that document answers (e.g. yes/no) or ratings (e.g. along a five-point scale from 'strongly agree' to 'strongly disagree') according to formats pre-specified by the author(s). Once all responses have been documented, they are each assigned numbers on the basis of a scoring procedure also devised by the author(s). These numbers are then added, averaged or otherwise transformed to generate individual scores on the measure. Many psychometric measures also supply population means that then facilitate interpretations of individual scores by systematic comparison with the scores of other people. However, the close reliance on self-reports is widely recognised, even within the psychometrics literature, to be associated with a range of interlocking problems, including those associated with the unreliability of introspection and memory, participant reactivity, response bias and demand characteristics (Rust and Golombok, 1999).

Second, psychometric measurement involves quantification: the translation of lived, human, experiential capacities, preferences, desires, opinions, beliefs, habits and characteristics into numerical indices purported to represent and enumerate them. The superficial reasonability of this – for a contemporary citizenry thoroughly schooled in the mores of a psychologised culture – can divert attention from the profound question of whether such attributes and capacities can actually *be* meaningfully quantified. Michell (2000) shows how, if it is to be considered properly 'scientific', psychometric quantification implicates a necessary sequence of two interdependent tasks. The first 'scientific' task 'involves devising test situations that are differentially sensitive to the presence or absence of quantitative structure' (Michell, 2000: 649). The second 'instrumental' task is one of generating and validating measures to reliably assess this attribute. This second task is only justifiable if the tests conducted during the first scientific task showed that the attribute being investigated is actually quantitative in character. For example, we are only justified in quantifying self-esteem using the Rosenberg Self-Esteem Scale (as did Packard *et al.*) if it is already established that self-esteem actually varies in accord with the systematic intervals such scales presuppose. Michell (2000) observes that whilst descriptions of the development of psychometric measures frequently present detailed accounts of the instrumental task, they never engage with the scientific task that necessarily precedes this work. Psychologists, then, are in the peculiar business of quantifying phenomena whose very character may not be quantitative at all.

Third, this raises the question of what is actually happening when participants complete psychometric assessments of personality or self-esteem. Packard et al., (2012) assessed three personality constructs (extraversion, neuroticism and psychoticism) using the Eysenck Personality Questionnaire – Revised

(EPQ-R). Rosenbaum and Valsiner (2011) investigated what such measurement actually entails using a well-regarded personality assessment that expands upon the EPQ-R and is often seen as its successor: the Neuroticism-Extraversion-Openness Personality Inventory, or NEO(PI) (Costa and McCrae, 1992). The NEO(PI) presents statements such as 'It doesn't embarrass me too much if people ridicule and tease me' and 'I have a very active imagination'. In relation to each statement, participants must rate themselves on a five-point scale from 'false' to 'true'. In accord with the scoring procedure, these ratings are then allocated numbers that represent loadings on one or more of the five personality constructs the measure assesses (openness to experience, conscientiousness, extraversion, agreeableness and neuroticism).

To reveal what occurs during psychometric assessment, Rosenbaum and Valsiner devised a novel procedure for administering the NEO(PI) that involved 'slowing down the rating process to open up the black box at its center' (2011: 61). Before marking the five-point scale from 'false' to 'true', the participants (students in both Estonia and the USA) were asked to write down, for each statement, what 'false' and 'true' meant. They were also asked to write down, for each statement, their rationale for choosing the scale point they endorsed. This procedure showed that the meanings of 'false' and 'true' differed, sometimes markedly, not only between participants but also from statement to statement. It further demonstrated that 'the same' answer was often underpinned by quite different rationales. The authors argued from these findings that, rather than objective measurement, psychometric testing actually involves the co-construction (between participant and measure) of a temporary dialogical field of meaning within which both statements and possible responses get actively and differentially interpreted. They concluded that

> it is a misplaced assumption that participants have direct access to their response and that this response is static and can be represented as a mark along a line ... rating scale data, despite being statistically manipulated, should not (and indeed cannot) be thought of as objective.
>
> (Rosenbaum and Valsiner, 2011: 61)

Fourth, Harré (2002) also considers at length the actual practices entailed in psychometric quantification. He notes that it is widely acknowledged that, historically, psychology adopted many of its ontological and epistemological commitments from physics; in striving to be accepted as a science, the fledgling discipline modelled itself upon science *par excellence*. But, Harré argues, in so doing, psychology failed to also adopt the crucial distinction that physicists routinely make between apparatus and instruments.

An instrument is a measuring device whose properties change predictably under the direct causal influence of some external force. A thermometer, for example, measures temperature by indicating changes in the volume of a quantity of mercury caused by its exposure to thermal energy. The more energy (the warmer it is), the greater the volume, and the relation between energy and volume is both directly causal and invariant – all other things (e.g. pressure) being equal. Apparatus, by contrast, is not designed to measure anything: its function is to *model* a system or phenomenon. Harré gives the example of a gas discharge tube wherein various gaseous compounds and their exposure to electrical charges can simulate the dynamics of the upper atmosphere. Modelling with apparatus can be extremely valuable, generating insights into situations where experimental or observational measures are difficult or impossible. Modelling, however, is most definitely not measuring; instruments measure, but apparatus merely models.

Psychometric questionnaires, Harré suggests, far more closely resemble apparatus than instruments. What actually occurs during their administration is not a process of measurement wherein direct causal forces impact upon respondents in a determinate, law-like fashion. Rather, it far more closely resembles a conversation with a psychologist, albeit a conversation that carefully and pre-emptively codes, structures and restricts the considerable variation that would otherwise ensue (cf. Potter and Wetherell, 1987). From Harré's perspective, Packard *et al.* were not so much measuring personality and self-esteem as engaging their participants in peculiarly restrained, inflexible and artificially structured conversations about them. But if psychometric questionnaires are actually restrictive analogues of everyday conversations, they are best understood as apparatus, not instruments. So, whilst they can (partially, artificially, restrictively) *model* psychological processes, they cannot be meaningfully said to *measure* them.

Harré notes that once we understand psychometric questionnaires as apparatus that models, rather than instruments that measure, this allows us to account for both their face validity (the fact that people are not perplexed or baffled when they try to answer them) and their reliability. He suggests that both face validity and reliability arise because the restrictive conversational modelling that occurs when psychometric questionnaires are administered necessarily relies upon exactly the same semantic rules and autobiographical and narrative conventions as everyday conversation. Since these rules and conventions are shared within (sub-)cultures, questionnaires that invoke them can readily be answered (face validity). Consistency (reliability) therefore reflects people's shared orientations to these meanings and conventions, just as variation reflects the degrees of freedom these conventions bestow.

Fifth, in psychometric measurement, the overall direction is one of increasing decontextualisation, abstraction and numerical sedimentation (Tolman, 1994). Participants must generalise from the many actual, shifting, embodied, material-spatial and dynamically contingent actualities of the ineffable fullness of a life being lived in order to generate fixed, precoded answers to decontextualised, standardised questions. These answers, already very partial, tenuous, static abstractions from the complex, fluctuating particularities of life itself, then get scored (according to the specified procedural rules). This yields numbers that are, in effect, still higher-level abstractions. In the Packard *et al.* study, these include numbers for 'self esteem' and for various personality traits. And in many studies (again, including Packard *et al.*), these higher-level abstractions themselves get aggregated or averaged across groups; the previously abstracted abstractions get abstracted still further, producing newer, even more nebulous abstractions, yet further removed from the lived particularity, concrete specificity and richness of any actual person or experience.

Sixth, and finally, publishing practice in psychology (and elsewhere) endorses the reporting of means and standard deviations of psychometric (and other) measures along with reports of statistically significant differences in these means between groups. For example, one of Packard *et al.*'s headline findings was that certain personality characteristics were statistically associated with unhealthy behaviours in their deprived group but not in their affluent group. But, as Billig (2013) explains, this reporting practice actually prevents us from knowing how many actual individuals in each group were affected by the reported difference, or to what extent. Even when exact numbers of participants are known, there is no straightforward way of working backwards from group means and probability levels to discern actual numbers of people whose attributes or actions differed from each other. To be clear, this is standard reporting practice and there is no imputation of immoral behaviour, whether by Packard *et al.* or by anyone else. Nevertheless, as Billig observes, a frequent effect of this practice is to facilitate publication of studies that find only relatively trivial differences between groups with respect to the variable of concern.

To summarise thus far: despite superficial appearances, the capacity of psychometric tests to quantify human experience and attributes has never actually been established. Indeed, it is possible that what occurs in psychometric testing is not a process of measurement at all; it is, rather, a process of modelling everyday conversation. Insofar as this modelling reproduces and accords with everyday semantic and procedural conventions, it can (imperfectly) reflect how variation in these conventions follows the contours of gross sociological or epidemiological differences such as levels of material deprivation. This distorted and partial conversational tracing, shaped at the sociological level, then gets mistaken for precision measurement at the psychological level. Moreover, to the extent that this process even begins to addresses the fluid, situated dynamism of individual lived experience, it necessarily decontextualises it, artificially rendering it abstract and static. The cumulative effects of these various shortcomings are nevertheless obscured by analytical, publishing and reporting conventions that prioritise group means, standard deviations and significance levels over numbers of actual cases or descriptions of the extent and ways in which people differ. These are general points about psychometrics, not specific to the study reported by Packard *et al.*; although as we have suggested, they are all relevant to that study. Their ongoing significances therefore need to be considered as we now discuss some of its details.

Psychological governance and health behaviours

Packard et al. (2012) claimed to have demonstrated an interaction between personality traits, mental wellbeing, health behaviour and social deprivation. To do so, they recruited 666 participants from areas of both high and low social deprivation, as assessed by the Scottish Index of Multiple Deprivation (SIMD): a measure produced by the Scottish Government that uses census data to allocate all of Scotland to one of 6,605 'datazones', each containing between 500 and 1,000 households. The SIMD then assigns each datazone to a ranking from 'most deprived' to 'least deprived', facilitating estimations of coarsely shared levels of deprivation for relatively small groups of residents in different areas of the country. Participants were administered a battery of self-report psychometric instruments on two separate occasions. First, they were asked to complete a set of health and behaviour assessments including a 'lifestyle' questionnaire (asking about exercise, smoking, diet, alcohol consumption, etc.) and a set of measures of wellbeing (the General Health Questionnaire, the Generalised Self-Efficacy Scale, the Sense of Coherence Scale, and the Beck Hopelessness Scale). Then, on a separate occasion two weeks later, they were asked to complete the Rosenberg Self-Esteem Scale and the EPQ-R.

Their main findings echoed similar claims from other studies: areas of higher social deprivation had higher levels of neuroticism and psychoticism, lower levels of wellbeing, and a greater prevalence of 'harmful health behaviours' such as poor diet, smoking and lack of exercise. In addition, though, Packard et al. report the

> novel finding ... that personality traits appeared to have a significantly greater impact on mental wellbeing among participants from more deprived circumstances. Further, personality and wellbeing impacted more on the pattern of health behaviours in this group compared with their more affluent counterparts.
>
> (2012: 620)

Specifically, amongst the most deprived group (but *not* amongst the least deprived), high extraversion and wellbeing scores significantly predicted greater consumption of fruit and vegetables; smoking cessation was predicted by high 'sense of coherence' and self-efficacy scores; and aerobic exercise engagement was significantly associated with high extraversion, self esteem and sense of coherence. We now look more closely at this study by exploring some statistical, psychometric and conceptual concerns.

Statistical concerns

Whilst 666 participants took part in this study, 2,712 were invited to do so; hence, the overall response rate was only 24.6 per cent. Moreover, this rate was markedly skewed between the less deprived (33.9 per cent) and the more deprived (19.0 per cent) areas. Whilst there is no universal definition of what counts as an acceptable response rate, higher rates are advisable where the intention – as in this study – is to generalise to the wider population. An overall rate of 24.6 per cent would be considered disappointing by many researchers and greatly increases the likelihood of recruitment biases in the sample that would make it unrepresentative. With particular respect to their personal qualities, it is well established in psychology that those who volunteer to take part in research differ significantly from those who do not (Rosenthal, 1965; Rosenthal and Rosnow, 1975), being judged on average to be more intelligent, more sociable, and more in need of the approval of others. The relatively low response rate in such a study makes it impossible to be confident that the findings can be reliably generalised without significantly more work to establish whether those who took part differed systematically from those who did not.

The statistical analyses controlled for a range of variables and interactions and seem to have been applied appropriately according to whether categorical or continuous variables were being investigated. It is also clear that many comparisons were conducted and that this allowed the authors to present a thorough analysis of their data. However, whilst reported significance levels were frequently high (as would be expected with a sample of this size), the authors do not say that they applied any statistical correction for the many comparisons they report, nor for the many others they almost certainly conducted but did not report. To illustrate: suppose the authors reported as significant any associations with a probability of less than 0.05. By definition, this significance level means that 1 in 20 apparently meaningful associations will in fact arise purely by chance. If more than 20 comparisons are made, this makes it almost certain that some of the associations discovered will be spurious whilst, simultaneously, making it impossible to know which. Whereas some of the reported significances in Packard et al. (2012) are as high as 0.00001 (one in 10,000) and so quite likely to have survived any correction, others are as low as 0.040 (one in 25). Moreover, the trend is for these lower significances to be associated with the 'novel' findings. In the absence of any correction for multiple testing, then, the veracity of these novel findings is questionable. In any case, as Billig's (2013) discussion of means and significance levels demonstrates, it is impossible to know just what these significance levels mean psychologically; for example, whether a lot of people in each group consistently differed just a little from each other or whether smaller numbers in each group differed greatly.

In addition, Packard *et al.* acknowledge that 'as a cross-sectional study it is not possible to infer causality from the observed statistical relationships' (p. 622). Hence, when they claim that personality traits predict wellbeing and health behaviours, the word 'predict' is meant purely in the statistical sense of describing the findings of a linear regression analysis. Prediction, here, is a function of the order in which variables are entered into an equation, not of real, longitudinal differences in actual data. In other words, and like many other psychological studies, Packard *et al.* obtained these results by first presuming that personality is more or less stable and that its constitution precedes or is more enduring than the health behaviours it is then assumed to predict. This presumption highlights a tension running throughout their paper with regard to personality because at various points, as here, it is treated as relatively stable; elsewhere, however, it is treated as more malleable. We return to this point below.

In any case, the absence of both longitudinal data and any experimental manipulation of variables makes it impossible to exclude the possibility that a third factor influenced both personality and health behaviour scores. In their research, the authors treat 'deprivation as a moderator variable' (p. 617): a variable that influences the strength of the relationship between two other variables. So even if we were to accept the claim that personality traits influence health behaviours, we would still not know how much of this influence was due to deprivation. In this context, it is important to recall that personality traits (extraversion, sense of coherence, self-esteem and self-efficacy) predicted wellbeing and health behaviours *only* amongst the most deprived participants; no such associations were found amongst the least deprived.

Psychometric concerns

The psychometric questionnaires used by Packard *et al.* were dependent on self-reports, and we have shown how this opens them up to a series of questions regarding their actual capacity to measure psychological states and processes. To further explore these issues, we now focus on the personality questionnaire used in this study: the 100-item EPQ-R. As the 12 sample questions presented in Table 5.1 illustrate, there is considerable potential for confounds *within the questionnaire itself* between the presumed personality characteristics the EPQ-R purports to measure and the socio-economic, material and health consequences with which this study associates them.

Quite clearly, people living in more deprived areas are less likely to have money available to donate to charity, to save or to purchase insurance, just as they are less likely to be able to engage in tax avoidance. Simultaneously, they are not only more likely to have debts, they are also more likely to worry about being indebted since, on average, they will find repayment more difficult. Epidemiologically, they are more likely to be afflicted with ill health (Wilkinson and Pickett, 2009). Also, by comparison with the more affluent, poorer people are more exposed to risks of assault, theft and burglary (Mirowsky and Ross, 1983); in circumstances like these, worrying more about health or about 'awful things that might happen' is almost inevitable. Similarly, people in more deprived areas will, on average, have fewer resources and less control over their material circumstances than those in more affluent areas and may be less liable to find the spontaneous warmth and kindness

Table 5.1 Sample questions from the EPQ-R

3. Does your mood often go up and down?
7. Would being in debt worry you?
8. Do you ever feel 'just miserable' for no reason?
9. Do you give money to charities?
34. Do you have enemies who want to harm you?
43. Do you worry about awful things that might happen?
52. Do you worry about your health?
75. Do you think people spend too much time safeguarding their future with savings and insurance?
77. Would you dodge paying taxes if you were sure you could never be found out?
83. Do you suffer from 'nerves'?
91. Would you like other people to be afraid of you?
95. Do people tell you a lot of lies?

Source: Packard et al., 2012.

from their neighbours that can help rebalance the scales of wellbeing (Clark and Heath, 2015). Consequently, they may well be more likely to find that their mood goes 'up and down' or that they sometimes feel 'just miserable' for no immediately apparent reason. Likewise, people in deprived areas more frequently, more continuously and more intimately encounter the threatening signs of their own deprivation: litter, graffiti, vandalism and derelict buildings, street drinking, drug use and visible gangs (Ross *et al.*, 2001). In such environments, increased prevalence of having 'nerves', of wishing to inspire fear in others, of being mindful of 'enemies' and sensitive to their possible 'lies' is to be expected simply because it is, at least to some degree, adaptive.

Packard *et al.* might retort that the EPQ-R assesses psychological variation that exists independently of such constellations of social and material circumstance, their associated power relations and unequal distributions of resources and opportunities, and the many psychological consequences of these. They might claim that by using the SIMD to recruit participants living in both deprived and affluent areas, their analyses expose a layer of psychological influence operating independently of these factors. Such claims would, however, be difficult to justify because, despite its use of multiple indicators, the SIMD allocates participants on the basis of living in (small) geographical areas – not on the basis of individual circumstances and the many interacting, changing complexities they encompass. The SIMD is an excellent resource that is entirely adequate for many public health purposes; nevertheless, since it clusters together groups of up to 1,000 households within each of its datazones, it operates at a level of granularity significantly coarser than would be required to warrant individualised psychological claims of this kind.

Conceptual concerns

This study raises various conceptual issues, many of which were alluded to in our general discussion of psychometric measurement. Here, we focus on the conceptual issues raised by the study's emphasis on personality. Whilst common-sense or folk psychological understandings typically assign considerable influence to individual personality characteristics, social psychologists demonstrated long ago that situational factors are frequently more influential (Mischell, 1973). Our primary concern is with the fundamental question of whether personality is an 'individual' attribute, somehow separable from environments and experiences, or whether it is an emergent accomplishment that incorporates and reflects the environments and experiences by which it is forged. As already noted, Packard *et al.* adopt conflicting stances to this question. When they treat personality as a predictor variable, they necessarily presume its relative stability and transcendent separation from circumstance. However, in their subsequent discussion, they cite a paper by Roberts *et al.* (2006) to support the contention that 'personality traits do change and may be modifiable' (cited in Packard *et al.*, p. 622). Relatedly, the personality construct 'sense of coherence' is represented by Packard *et al.* as a relatively stable and individual difference, even though Antonovsky (1993), whom they also cite, describes it as a pervasive and enduring feeling of confidence arising a) when the stimuli from one's internal and external environments in the course of living are structured, predictable and explicable; b) resources are available to one to meet the demands posed by these stimuli, and; c) these demands are challenges worthy of investment and engagement.

So Packard *et al.* seem to want it both ways: they want personality to be individual and relatively stable, but they also want it to be situational and relatively malleable. Whilst this conceptual duality seems necessary to warrant the paper's key recommendation that health interventions may be more effective when adapted to certain personality characteristics, it simultaneously exposes an ideological position that simply accepts 'personality' as an individualised psychological construct. Sloan (1996) thoroughly critiques mainstream personality psychology, demonstrating how it typically operates in the service of a deeply unequal societal status quo. Personality psychology reflects and stabilises the norms and precepts of dominant social orders within particular historical periods, whilst its claims to generalisability and universality simultaneously obscure these associations. In doing so, it emphasises individuals, so deflecting attention from the continuous social and material shaping of experience and behaviour. This was clearly illustrated in the way the EPQ-R could code reasonable responses to deprivation and inequality as enduring facets of individual personalities.

In sum, the six general concerns that we identified above are supplemented in Packard *et al.*'s study by a further series of statistical, psychometric and conceptual concerns that simultaneously realise and obscure their relevance. Once examined more closely, the 'novel' associations with personality characteristics that they reported are less solid than they seem; the questionnaire used to identify most of these features is unable to clearly distinguish personal predilections from reasonable responses to deprivation; and (perhaps unsurprisingly) the study's notion of personality is therefore both conceptually confused and ideologically shaped.

Biopolitics and psychological governance

Understood as practical biopolitics, rather than as objective, value-free, morally neutral science, psychology – as exemplified by the study discussed here – provides an ostensible truth about life: personality variables predict unhealthy behaviour amongst poorer people but not amongst richer ones. On the basis of this ostensible truth, it legitimates policies designed to impel poorer people to live in certain ways. As noted, Packard *et al.* recommend that health promotion activities should pay more attention to individual personality traits. This recommendation dovetails neatly with other policy suggestions that individualise health and make it a matter of personal responsibility, including some recent BIT initiatives, Department of Health recommendations regarding health promotion (2010) or which emphasise and promote personal choice within health services (2015), and health service strategies of limiting or rationing access to surgery by obese people and smokers (Donnelly, 2015). Such strategies and policy suggestions legitimate and normalise the link between the good and responsible life they presume and the avoidance of those unhealthy behaviours that the irresponsible poor stubbornly persist in favouring.

At the extreme, then (and we recognise that this is unlikely to have been Packard *et al.*'s intention), the study might even be read as providing implicit justification for the biopolitical power to let die. If personality really were individual, fixed and relatively stable, there may be a sense in which poor people endowed with traits that predispose them to live unhealthily can be viably seen as simply beyond the reach of health and welfare initiatives. Once these people are cast as individual, irrational agents who freely persist in making unhealthy choices, it seems quite clear where responsibility lies. So, with this biopolitical focus, Packard *et al.*'s paper reproduces neoliberal tropes of individualism, responsibility and choice whilst simultaneously recommending governmental techniques and practices that accord with them. In this regard, we note with dismay the evidence that annual death rates in the UK, which had been declining since the 1970s, began rising again in 2011 and jumped sharply upwards by 5.4 per cent in 2015 compared to the previous year – an extra 27,000 deaths (West, 2016).

Notwithstanding Packard *et al.*'s neoliberal emphasis on individual personality factors, it might be argued that their proposals for action represent an advance on individual therapeutic approaches that would change the contents of people's heads using education, cognitive restructuring and behavioural exercises. By contrast, Packard *et al.* suggest that health promoters might seek to change thinking and behaviour by altering the immediate contexts in which people live 'through the manipulation of resources and experiences that influence cognitions that underlie some of the measures of wellbeing reported in this study' (p. 622). Although they do not articulate how these aims could be achieved, presumably they are referring to community initiatives such as support groups, educational campaigns and enhanced recreational opportunities. However, the assumption that significant improvements in wellbeing can be achieved this way amounts to what Epstein (1993) calls 'social efficiency': the notion that there are cheap, straightforward, personal solutions to enduring social and moral problems. The reality, though, is that the basic pattern of health inequalities has a disturbing persistence, even in 'advanced' countries like ours. Dorling (2013) describes how some districts of London have kept their sombre profiles of poor health and mortality across decades, even centuries, despite dramatic shifts in the ethnic composition and culture of the communities that occupy them: the 'longer people stay both in poverty and in poor places, the earlier they tend to die' (Dorling, 2013: 46).

Far from being the product of irrational or irresponsible choices impelled by individual personality characteristics, such analyses suggest that health inequalities are deeply embedded in the social and material fabric of the world and are best understood as an expression of it. One of the most high-profile pieces of research demonstrating this is the 'Whitehall study': an ongoing longitudinal survey of the health, workplace roles and lifestyles of four occupational grades of middle-class civil servants in London, enlisting 17,000 participants. It has shown that those in the top administrative grade suffered the fewest deaths over a decade, while those in the lowest grade suffered a threefold higher mortality rate. This gradient remained, even for the more health conscious who were non-smokers, exercised and took care of their physical condition; their risk of illness and early death dropped by no more than a third (Marmot, 2010). And, of course, even the lowest ranking of these groups of civil servants probably experienced more benign circumstances than the most deprived participants in Packard *et al.*'s study.

Taken together, recent epidemiological (Stuckler and Basu, 2013; Wilkinson and Pickett, 2009), ethnographic (McKenzie, 2015) and phenomenological (Charlesworth, 1999) studies suggest three things: first, that to be poor in a society of consumers is frequently to be marked as unworthy of respect – by others, and by one's self; second, that this knowledge itself makes a direct contribution to poor mental and perhaps even poor physical health; and third, that impoverishment can render 'unhealthy' habits all the more attractive because, amongst other things, they provide consolation and escape. As long ago as the 1930s, George Orwell described how the media were fond of stories berating working-class people for favouring unwholesome foods like condensed tinned milk, sugar and white bread. But Orwell saw that, far from being the insignia of moral turpitude, these preferences were understandable (even 'rational') responses to the predicaments in which so many found themselves:

A millionaire may enjoy breakfasting off orange juice and Ryvita biscuits; an unemployed man doesn't. When you are unemployed, which is to say when you are underfed, harassed, bored and miserable, you don't want to eat dull wholesome food. You want something a little bit 'tasty'.

(Orwell, 1937: 49–50)

In a similar vein, Graham (1987) showed how some women living in poor areas smoked *despite* being thoroughly cognisant of its adverse health consequences. For these women, smoking alleviated the strains of poverty, helping them seize time for themselves and better manage the demands they faced:

I smoke when I'm sitting down, having a cup of coffee. It's part and parcel of resting.

I think it gives me a break. Having a cigarette is an excuse to stop for five minutes.

Sometimes I put him [child] outside the room, shut the door and put the radio on full blast and I've sat down and had a cigarette, calmed down and fetched him in again.

(Graham, 1987: 52–4)

Like Orwell's ethnography, Graham's study demonstrates that there are situations in which 'unhealthy' behaviours are best understood as quite rational responses to enduring combinations of toxic social and material circumstances. This suggests that psychology in general, and so-called individual differences in personality and behaviour in particular, simply cannot be understood apart from the circumstances in which they arise; personality and behaviour are continuously and profoundly shaped by ongoing interactions between individual and environment. This implies that if researchers are serious about explaining and challenging health inequalities, their focus should very much include relevant social issues and material influences.

Implications for research, policy and practice

Research would therefore be more effective in genuinely making a difference to health inequalities if studies of personality and behaviour were to *simultaneously* address issues including (but not restricted to) the following:

- the dominance of individual psychotherapy models that locate clinical distress as 'internal' and largely disconnected from people's history and current circumstances;
- the recruitment of psychology into policing how people live within increasingly narrowed parameters of what is deemed 'normal';
- the commodification of food and the global trading of food staples such as wheat and rice;

- the dominance of food corporations and their failures of ethical responsibility with regard to wider communities, both local and global;
- the hollowing out of local high streets, reducing peoples' options around what and where they can buy foods;
- the stripping of local assets leading to the closures of swimming pools, leisure centres, parks, playing fields and other resources that support exercise;
- the marketing, media and advertising practices that influence peoples' choices around food;
- the ineptitude of governments who fail to hold corporations to account, instead preferring voluntary 'responsibility deals';
- increasing poverty, leading to the increased use of food banks – not because people are opportunistically more aware of them but because of forced choices between basic essentials to support life;
- the reduction and removal of opportunities for skill sharing between generations, dressed up as 'progress', leaving people without knowledge, time or opportunity to do things such as cook for themselves and forcing increased reliance on processed and prepared foods;
- the health consequences of living in overcrowded, substandard housing;
- the promotion of competition, as opposed to collaboration or solidarity, that encourages people to emphasise destructive comparisons between each other;
- the damning language of recent years that has increasingly sought to shift responsibility onto individuals for their predicaments, and the associated illusion that all can be made better simply by increased individual effort.

It is by sidelining research into topics such as these and promoting instead a research agenda narrowly focused upon individuals and their presumed individual characteristics that psychology functions as a practical biopolitics in the service of a toxic neoliberal governmentality. Here, psychologists might learn from colleagues in consanguineous disciplines who have critiqued policies (Friedli, 2013), empirically demonstrated their consequences for health services and workers (Crawshaw, 2013), and provided alternative conceptualisations (such as a focus on health practices rather than health behaviours – see Cohn, 2014) and have, in these ways, generated research considerably more capable of addressing these problems.

Finally, with respect to policy and practice, we should highlight the relevance of initiatives developed under the aegis of community psychology (Kagan *et al.*, 2011; Orford, 2008) that take seriously and work with the contextual, social and material determinants of health and wellbeing. In a similar vein, we welcome recommendations for social prescribing: the delivery and facilitation within primary care of 'opportunities for arts and creativity, physical activity, learning new skills, volunteering, mutual aid, befriending and self-help, as well as support with, for example, employment, benefits, housing, debt, legal advice, or parenting problems' (Friedli *et al.*, 2009: 3). Simultaneously, though, we take seriously Epstein's (1993) warning regarding the transient or even illusory character of the benefits to be gained by tinkering with relatively minor aspects of people's circumstances. With this in mind, we call also for fundamental and widespread changes to social and economic policy – changes that would substantially redress inequality and, in so doing, improve the health and wellbeing of us all.

References

Antonovsky, A. (1993) 'The structure and properties of the sense of coherence scale'. *Social Science and Medicine*, 36(6): 725–733.

Behavioural Insights Team (2015) 'Who we are' [online]. *Behavioural Insights Team*. Available at: www.behaviouralinsights.co.uk/about-us/

Billig, M. (2013) *Learn to Write Badly: How to Succeed in the Social Sciences.* Cambridge University Press, Cambridge.

Centre for Economic Performance's Mental Health Policy Group (2006) The Depression Report: A New Deal for Depression and Anxiety Disorders. CEPOP15. Centre for Economic Performance, London School of Economics and Political Science, London. Available at: http://eprints.lse.ac.uk/818/

Charlesworth, S. (1999) *A Phenomenology of Working Class Experience*. Cambridge University Press, Cambridge.

Clark, T. and Heath, A. (2015) *Hard Times: The Divisive Toll of the Economic Slump*, second edition. Yale University Press, New Haven and London.

Cohn, S. (2014) 'From health behaviours to health practices: an introduction'. *Sociology of Health and Illness*, 36(2): 157–162.

Costa, P. and McCrae, R. (1992) *NEO PI-R Professional Manual*. Psychological Assessment Resources Inc., Odessa, FL.

Crawshaw, P. (2013) 'Public health policy and the behavioural turn: the case of social Marketing'. *Critical Social Policy*, 33(4): 616–637.

Cromby, J. and Willis, M. E. H. (2014) 'Nudging into subjectification: governmentality and psychometrics'. *Critical Social Policy*, 34(2): 241–259.

Deleuze, G. (1992) 'Postscript on the Societies of Control'. *October*, 58 (Winter): 3–7.

Department Of Health (2010) Healthy Lives, Healthy People: Our Strategy for Public Health in England. CM7985. Department of Health, London. Available at: https://www.gov.uk/government/uploads/system/uploads/attachment_data/file/216096/dh_127 424.pdfDepartment Of Health (2016) 'NHS Choice Framework' [online]. *Department of Health*, 29 April. Available at: https://www.gov.uk/government/uploads/system/up loads/attachment_data/file/417057/Choice_Framework_2015-16.pdf

Donnelly, L. (2015) 'Most NHS areas refuse surgery for obese patients'. *The Daily Telegraph*, 6 March. Available at: www.telegraph.co.uk/journalists/laura-donnelly/ 11454496/Most-NHS-areas-refuse-surgery-for-obese-patients.html   (accessed   11 February 2016).

Dorling, D. (2013) *Unequal Health: The Scandal of Our Times*. Policy Press, Bristol.

Epstein, W. (1993) *The Dilemma of American Social Welfare*. Transaction, New Brunswick, NJ.

Foucault, M. (2008) *The Birth of Biopolitics: Lectures at the Collège de France, 1978–1979* (trans. G. Burchell). Palgrave Macmillan, Basingstoke.

Friedli, L. (2013) '"What we've tried, hasn't worked": the politics of assets based public Health'. *Critical Public Health*, 23(2): 131–145.

Friedli, L., Jackson, C., Abernethy, H. and Stansfield, J. (2009) Social Prescribing for Mental Health: A Guide for Commissioning and Delivery. Care Services Improvement Partnership, North West Development Centre, Manchester. Available at: www.centreforwelfarereform.org/uploads/attachment/339/social-prescribing-for-mental-health.pdf

Graham, H. (1987) 'Women's smoking and family health'. *Social Science and Medicine*, 25(1): 47–56.

Harré, R. (2002) *Cognitive Science: A Philosophical Introduction*. Sage, London.

Kagan, C., Burton, M. and Duckett, P. S. (2011) *Critical Community Psychology: Critical Action and Social Change*. Wiley, London.

Lemke, T. (2016) 'Rethinking Biopolitics', in S. Wilmer and A. Zukauskaite (Eds.) *Resisting Biopolitics: Philosophical, Political and Performative Strategies*. Routledge, New York, pp. 57–73.

McKenzie, L. (2015) *Getting By: Estates, Class and Culture in Contemporary Britain*. Policy Press, Bristol.

Marmot, M. (2010) Fair Society, Healthy Lives: The Marmot Review. UCL Institute of Health Equity, London. Available at: www.instituteofhealthequity.org/projects/fair-so cietyhealthy-lives-the-marmot-reviewMichell, J. (2000) 'Normal science, pathological science and psychometrics'. *Theory and Psychology*, 10(5): 639–667.

Mindfulness All-Party Parliamentary Group (2015) Mindful Nation UK: Report by the Mindfulness All-Party Parliamentary Group (MAPPG). MAPPG, London.

Mirowsky, J. and Ross, C. E. (1983) 'Paranoia and the structure of powerlessness'. *American Sociological Review*, 48(2), 228–239.

Mischell, W. (1973) 'Toward a cognitive social learning reconceptualization of Personality'. *Psychological Review*, 80(4): 252–283.

Moloney, P. (2013) *The Therapy Industry: The Irresistible Rise of the Talking Cure, and Why it Doesn't Work*. Pluto Press, London.

New Economics Foundation (2004) A Well-being Manifesto for a Flourishing Society. New Economics Foundation: London. Available at: www.neweconomics.org/publica tions/entry/a-well-being-manifesto-for-a-flourishing-society

Orford, J. (2008) *Community Psychology: Challenges, Controversies and Emerging Consensus*. Wiley, London.

Orwell, G.. (1937) *The Road To Wigan Pier*. Gollancz, London.

Packard, C. J., Cavanagh, J., McLean, J. S., McConnachie, A., Messow, C.-M., Batty, G. D., Burns, H., Deans, K. A., Sattar, N., Shiels, P. G., Velupillai, Y. N., Tannahill, C. and Millar, K. (2012) 'Interaction of personality traits with social deprivation in determining mental wellbeing and health behaviours'. *Journal of Public Health*, 34(4): 615–624.

Potter, J. and Wetherell, M. (1987) *Discourse and Social Psychology: Beyond Attitudes and Behaviour*. Sage Publications, London.

Pressman, S. and Cohen, S. (2005) 'Does positive affect influence health?'. *Psychological Bulletin*, 131(5): 925–971.

Roberts, B. W., Walton, K. E. and Viechtbauer, W. (2006) 'Patterns of mean-level change in personality traits across the life course: a meta-analysis of longitudinal studies'. *Psychological Bulletin*, 132(1): 1–25.

Rose, N. (1985) *The Psychological Complex*. Routledge, London.

Rosenbaum, P. and Valsiner, J. (2011) 'The un-making of a method: from rating scales to the study of psychological processes'. *Theory and Psychology*, 21(1): 47–65.

Rosenthal, R. (1965) 'The volunteer subject'. *Human Relations*, 18(4): 389–406.

Rosenthal, R. and Rosnow, R. (1975) *The Volunteer Subject*. Wiley, New York.

Ross, C. E., Mirowsky, J. and Pribesh, S. (2001) 'Powerlessness and the amplification of threat: neighbourhood disadvantage, disorder and mistrust'. *American Sociological Review*, 66(4): 568–591.

Rust, J. and Golombok, S. (1999) *Modery Psychometrics: The Science of Psychological Assessment*, second edition. Routledge, London.

Sloan, T. (1996) 'Theories of personality: ideology and beyond', in D. Fox and I. Prilletensky (Eds.) *Critical Psychology: An Introduction*. Sage, London, pp. 87–103.

Stephenson, N. and Papadopoulos, D. (2007) *Analysing Everyday Experience: Social Research and Political Change*. Palgrave Macmillan, London.

Stuckler, D. and Basu, S. (2013) *The Body Economic: Why Austerity Kills*. Allen Lane, London.

Tolman, C. (1994) *Psychology, Society, Subjectivity: An Introduction to German Critical Psychology*. Routledge, London.

West, D. (2016) 'Exclusive: rocketing death rate provokes calls for national investigation' [online]. *Health Services Journal*. Available at: www.hsj.co.uk/topics/policy-and-regulation/exclusive-rocketing-death-rate-provokes-calls-for-national-investigation/7002408.fullarticle#.VsIGuqG__uA.twitter (accessed 15 February 2016).

Wilkinson, R. and Pickett, K. (2009) *The Spirit Level: Why Equality is Better for Everyone*. Penguin, London.

6 'What about the children?' Re-engineering citizens of the future

Val Gillies and Rosalind Edwards

Introduction

Debates in policy and practice are often oblivious to the replication of similar themes and solutions through the centuries. The idea that deprivation is transmitted through the generations via the mind of the child is particularly influential. The belief that early childhood experiences profoundly shape the personality, behaviour and destiny of individuals has exerted a potent allure across time, mutating and adapting to fit the political and cultural contours of the day (Kagan, 1998). Interventions focusing specifically on the interior of the child can be traced back to a religious reclamation of young souls from the dissolute poor, in effect promoting virtues of productive citizenship rooted in economic liberalism. Attention then shifted from the soul to the body and mind in the nineteenth century as psychological theories of child development emerged, eventually to endorse embedded liberal normativities through psycho-analytic models of family functioning. Current incarnations of infant determinism are conveyed through the language of cutting-edge brain science with emphasis placed on new discoveries and the transformative potential unlocked by such knowledge (Edwards *et al.*, 2015). New morally infused prescriptions for family relationships have followed, inspiring legislative change and state-enforced coercion. Characterising each of these eras is an institutionalised effort to initiate behaviour modification in the name of prevention.

In this chapter, we explore the history of ideas about intervention in family, highlighting attempts to (re)engineer children's upbringing for the sake of the nation's future. We show how the specific goals of personal and psychological governance shift over the centuries to reflect politically grounded representations of the national good. In adopting this long-term perspective, we aim to unsettle present-centred assumptions and reveal the way that social concerns and psychologically directed remedies play out through extended cyclical patterns. In line with this approach, psychological governance is broadly defined as attempts to advance, manage and regulate the social good through targeting the minds of individuals as a means of changing their behaviour. We consider the relationship between programmes and activities designed to address social dis-ease (poverty, crime and disorder) and understandings of the role of parents in the context of shifting emphases of political systems across time. We detail how nineteenth-century concerns about children's moral development gave way to a preoccupation with their physical health and genetic heredity, which then transmuted into anxieties about their psychological development and, more latterly, the quality of infant neurological architecture. While the theorising of child psychology and related modes of family intervention shifts, a conviction remains that optimally formed minds and bodies can prosper within a capitalist system.

Saving the children

Familial accountability for the welfare and moral profile of offspring has long been assumed. But specific targeting of children and families for intervention can be traced back to nineteenth-century efforts to address the human suffering and social costs associated with laissez-faire liberal capitalism. Earnest conviction that free trade and the pursuit of self-interest upheld the best interests of all led the powerful and privileged to seek explanations for the misery and dysfunction that surrounded them. Deprivation and destitution were acute while crime and social disorder remained a constant threat, particularly in London where many of the wealthy elite resided. Victorian efforts at social reform were funnelled through the dominant conceptual framework of classical liberalism, implicating the dubious moral character of those struggling to survive. Hardship and privation were viewed as temporary wrinkles, in an otherwise benevolent system, that could be ironed out through strengthening the moral fibre of the nation (Rooff, 1972). Reflecting the tenets of liberalism, the state was to play a minimal role in managing those who were not managing themselves. Bolstered by the growing influence of evangelism, philanthropy took on a new significance as a way for the Victorian rich to understand and control the unwashed masses through the issuing of relief alongside moral surveillance and counsel (Stedman Jones, 2013).

As Rose (1999) notes, Christianity played a central role in fashioning modern Western conceptions of personhood, with the notion of a bounded soul laying the foundations for the development of psychological science. As religious experiences came to be located in the hearts and minds of individuals, this relationship with God was identified in the qualities of the self. As such, 'character' emerged as an early psychological model, heralding the influence of psychological discourses in shaping how we understand what a child 'is' and opening up mind and behaviour to public scrutiny, self-evaluation and redemption. Victorian philanthropists sought to inculcate self-governance as a moral sensibility antithetical to the social ills blighting their towns and cities. Poverty was approached as symptomatic of a lack of drive, resilience and self-respect. Worse still, pauperism and reliance on the state Poor Law marked a shameful lack of foresight and self-control.

The second half of the nineteenth century saw an unprecedented proliferation of private charities and relief agencies and, significantly, the singling out of children as the focus for moral ministry. Commonly represented by their benefactors as innocents who could be saved from the bad character, degradation and degeneracy that had befallen their parents, children began to feature prominently in a range of social and religious associations. Many contemporary children's charities including Barnardo's, Action for Children and The Children's Society have their roots in what became known as the British child rescue movement. Propelled by a heady romantic vision of childhood and an imperialist concern with purifying the race and strengthening the nation, these organisations succeeded in extending legal protection to children – though, crucially, through the depiction of an abusive debauched residuum routinely exploiting or abandoning their offspring. The broader social and structural context framing childhood and family experiences of deprivation and risk were overlooked for a sensationalist focus on the sins of the parents. Children of the poor were depicted as little barbarians; victims of their cruel, depraved parents and in need of assimilation into the ranks of respectable British society.

Thomas Barnardo is amongst the best known of the child rescuers, having set up the first of his homes for 'waifs and strays' at Stepney East London in 1870. Combining a flamboyant celebrity image with evangelical self-righteousness, Barnardo did considerably more than simply provide for orphans. He actively constructed the children of the poor as a category apart, embodying the savagery of the degenerate classes but also the potential for deliverance. As Lydia Murdoch (2006) outlines, Barnardo skilfully developed the equivalent of missionary conversion parables, drawing on popular melodramatic tropes to raise money and justify the removal of children from their families. At the Stepney institution, a photographic studio was installed in 1874 with images of 55,000 of Barnardo's charges captured. Striking 'before and after' publicity shots were produced from there as well as other fundraising material. To ensure the children looked convincingly neglected, many had their shoes removed, clothes deliberately torn, hair tangled and dirt smeared on their faces. The post-'rescue' photographs showed the children transformed by Christian educa-tion and honest toil, depicted as tidy, shiny-faced 'little workers' holding a broom or engaged in a useful trade.

Reflecting the imperialist sensibilities characterising Victorian Britain, the psychological plasticity attributed to children was viewed through a distinctly racialised lens. The children of the poor were identified as a risk to the heredi-tary superiority of British stock, but also as a malleable resource born with a kernel of racial superiority that could be nurtured or left to degenerate (Boucher, 2014). Child rescue narratives commonly drew on explicitly racialised imagery, describing missions to civilise little 'street Arabs', 'urban savages' or 'raggamuffin tribes' and depicting the children as 'specimens' with darkened skin and exaggerated facial features. According to Barnardo, this physiognomy could undergo complete 'metamorphosis' under the auspices of his training, marking the child's new-found purity and religious salvation (cited in Murdoch, 2006). This narrative thus posed a link between the psychological and the physical; a transformation of the soul became evident and evidenced in a transformed appearance.

The notion that the British character of the young poor could be developed to shore up world supremacy was not confined to the high-profile philanthropists of the day. Deep faith in the reformatory power of British children eventually drove a systematic migration programme, resulting in tens of thousands of the rescued 'waifs' being sent to underpopulated settler colonies in Australia, New Zealand, Canada and South Africa. The philanthropist Maria Rye was among the first to send regular parties of 'gutter children' to Canada to work as indentured servants, estimating that the 'expense of taking a child out of the gutters in London, and placing it in Canada … may be roughly reckoned at £15 per head' (Liverpool Maritime Archive and Library, 2014: 3). This practice of finding and transporting suitable child migrants was legislated for and part funded by the British state right up until the 1950s. The investment was regarded as mutually beneficial, delivering the children from wretchedness, demoralisation and temptation, while extending the reach and strength of the British Empire. By the late nineteenth century, over 50 charitable organisations were regularly dispatching poor children abroad, including The Church of England Waifs and Strays Society and National Children's Homes. However, the substantial numbers of children shipped out of the country were dwarfed by the numbers institutionalised ultimately because their families lacked the resources to care adequately for them.

Thomas Barnardo was by far the most enthusiastic and prolific perpetrator of what he termed 'philanthropic abduction'. This phrase hints at the harsh reality lying behind this systematic programme of salvation and the severing of familial and community ties it entailed. His campaign literature was filled with lurid accounts of violent tussles with drunken, bestial mothers determined to keep their neglected urchins as exploitable property. But while the philanthropists were at pains to represent their targets as abandoned, abused and unloved, most institutionalised children had been embedded in family networks. As Murdoch (2012) has demonstrated, parents often made considerable efforts to stay in touch with their children and monitor their welfare. Destitute relatives often temporarily placed children in charitable institutions while they got back on their feet, subsequently returning to find they had been sent away without any warning or notification.

The practice of isolating poor institutionalised children from the perceived bad influence of their family was vigorously defended by the child-centric philanthropists, often in court custody battles. Parents were portrayed as wicked, immoral and brutal and their children, as suffering 'worse than orphans' (Barnardo, 1885). Yet ideals of home, family and hearth continued to exert great influence over Victorian consciousness (Behlmer, 1998). Many competing philanthropists regarded 'the family' as a sacred wellspring of personal responsibility and British character, focusing their efforts on remoralizing poor families as a whole by strengthening their character and resilience. This family-centred approach to poverty was to eventually evolve into statutory social work practice.

In line with the cultural sensibilities of the day, the child rescuers also expressed great faith in the redemptive power of family love – though privileging an artificial operational framework of domesticity over any blood or community relations. For example, Barnardo established substitute family settings styled as 'family cottages'. Administered by 'foster mothers' and housing between 20 and 40 children, these 'cottages' and 'village homes' were located away from urban squalor and the corrupting influence of the adult poor. While a distinct contrast from the regimental conditions of the workhouse or state orphanage, the ideal of family that was promoted was largely reduced down to training in British character and moral citizenship (Swaine, 2011).

The science of reform: strengthening British stock

By the turn of the twentieth century, the fervour and conviction powering the 'philanthropic abduction' of poor children was waning. Severe and prolonged economic instability and the rise of socialist ideals saw the foundations for the welfare state laid, while the tenets of classical liberalism came under pro-longed attack. A key catalyst for this reformist agenda was the unexpected struggle faced by British forces in winning the Boer War (Kuchta, 2010). The conflict was expected to quickly establish the might of the British Empire, but it dragged on for years, provoking fears over 'national efficiency'. The appalling physical condition of young men in the army recruitment pool was quickly established, triggering panic about Britain's imperialist supremacy. Laissez-faire principles fell from favour as the state began to link the health of its children with national competitiveness.

The organisations established by the child rescuers continued running resi-dential children's homes but within a context that saw increasing involvement of the state in overseeing child welfare. As part of the Liberal government reforms, medical inspections of children were introduced in schools and free school meals were provided for the poor. Classes in 'mothercraft' were also founded to encourage the raising of fitter children through providing advice on feeding and physical care (Kent, 2002). By this point, the general consensus posi-tioned children as raw material to be shaped in the interests of the nation. This moved beyond the introduction of welfare reforms and a compulsory education system as attention began to converge around new scientific accounts of human development. In 1907, the Child Study Society was set up, followed by the medically orientated Childhood Society. The proponents of these organisations were informed by a range of circulating theories and ideas about childhood, not least influential debates about the social consequences of inheritance and bad breeding (Behlmer, 1998).

Conceptions of heredity and the laws of biology found particular resonance with the Social Darwinist instincts of the elite, inspiring a new generation of philanthropists and social reformers to rally around eugenic reasoning.

Founded in 1909, the Eugenics Education Society sought to inform the public about the principles of selective breeding while lobbying the government for controls on fertility. The new 'science' of eugenics had wide and broad appeal across the political spectrum of the establishment but proved particularly attractive to those on the left, including the Fabian founders of the Labour Party and those who considered themselves radical reformers (including feminists and Marxists). As Dikötter (1998) notes, between the two world wars, eugenics belonged to the political vocabulary of virtually every significant modernising force in the Western world. In comparison with the US and Scandinavia, Britain was among the more cautious adopters of eugenic legislation, but support for the prevention of inbreeding through segregation of 'defectives', 'inebriates' and those with venereal disease was passionately discussed in Parliament. In 1913, the Mental Deficiency Act was passed with relatively little opposition, allowing the compulsory detention of those deemed 'unfit' despite never having committed a crime or been certified.

The early advancement of British child psychology was grounded within this eugenic paradigm under the auspices of Cyril Burt and his associates. Particular emphasis was placed on the development of intelligence testing and the ranking and sorting of children. Indeed, after the passing of the Mental Deficiency Act, Burt was appointed by the London County Council as their official education psychologist with the aim of classifying children and weeding out the 'feeble-minded' for admission to schools for the mentally defective (Stewart, 2013). But this period also marked a more general interest in defining and policing the parameters of normal child development. Sensitised to any manifestations of abnormality, officials and charity agencies increasingly began to refer children to the new psychiatric professions. As concern shifted from the moral development of children to their physical health and then to the organically inscribed workings of their minds, the Maudsley psychiatric hospital in London was forced to set up a separate children's department to deal with the rapid growth in numbers of children being treated during the interwar years (Evans et al., 2008). Most of these referrals concerned children living in deprived conditions, many of them suffering from malnutrition and general poor health. Details collected about the child's physiology, habits, personality and family relationships were used to form diagnoses, primarily of physical or neurological abnormalities, 'moral disorder' or mental deficiency.

The psychological models drawn on to categorise and treat children grew in sophistication through cross-fertilisation with the new behaviourist and psychoanalytic models and, more specifically, the emergence of the child guidance movement in the US. This precipitated the expansion of psychological horizons beyond the constraints of abnormality and deficit to encompass the risk of 'maladjustment' faced by otherwise normal children. As John Stewart (2013: 14) outlines, child guidance proponents came to emphasise the preventative function in promoting emotional and psychological development through 'the dangerous age of childhood'. Efforts were to be focused on early indicators of disturbance, which included a broad range of behaviour such as bed-wetting, misconduct, shyness and other manifestations of nonconformity. These were viewed as symptoms of deeper dysfunctions rooted in the child's family relationships, ensuring normal development became imbued with a sense of fragility (Rose, 1999). But crucially, children were regarded as uniquely mouldable and responsive to treatment administered via advisory council to parents.

Again, philanthropy played a formative role in establishing child guidance clinics on both sides of the Atlantic. In the US, large philanthropic donations from, amongst others, Rockefeller and the Commonwealth Fund helped establish a network across the US to pursue the study, treatment and prevention of juvenile psychological disorders. Reflecting a more general post-war secularisation of philanthropic activity, moving away from principles informed by religion toward ideals of science, donors placed their faith in the power of therapy to uplift the human condition (Rosenberg, 2002). As Alice Smuts (2008) documents, the generosity of the philanthropists was underpinned by their conviction that shaping the mental, physical and moral development of the child was a way of controlling and directing their future and that of the nation. This imaginary was widespread and shared by many in positions of government. In 1930, President Herbert Hoover convened a White House conference with over 3,000 in attendance to discuss the issue of child health and wellbeing, declaring 'if we could have one generation of properly born, trained, educated and healthy children a thousand other problems of Government would vanish' (cited in Smuts, 2008: 4).

American philanthropy was similarly instrumental in sponsoring the child guidance movement abroad. Following requests from British advocates, the Commonwealth Fund extended finance to support exchange observation visits and the setting up of clinics, supplementing the smaller-scale investments of British philanthropists committed to further developing child mental health services (Stewart, 2013). After the Second World War, the child guidance system was expanded and institutionalised as part of the post-war reconstruction effort. Concern over the traumatic impact of the Blitz and evacuation on the psychological wellbeing of children led to demands for an integrated service that was to be administered through local education authorities. This was in a context where a vision of 'the family' was broadly promoted as an essential mechanism of reconciliation and order after an extended period of chaos and uncertainty (Thomson, 2013).

During this period, psychoanalytic accounts of child mental health rose to ascendancy, propelled by the psychiatrist John Bowlby's theorising around maternal deprivation and attachment. While working at the London Child Guidance Clinic during the war, Bowlby came to believe that the key to normal development was located in the warmth and consistency of mother–child relationships. Drawing on examples of clinic cases, he attributed the development of deviant personalities to maternal separation or poor-quality relational bonding (Stewart, 2013). Attachment theory found great resonance among those who positioned the traditional family as the essential civilising force driving the evolution of social democracy. Mounting anxiety about increasing divorce rates in the aftermath of the war and the incidence of child neurosis was offset by a broader optimistic conviction that state intervention had the capacity to solve all social problems (Shapira, 2015). As such, the emotional as well as the physical welfare of children became incorporated into the post-war settlement shaping the development of Keynesian social and economic policy.

As Mathew Thomson (2013) argues, Bowlby's attachment theory was pivotal to the development of a limited welfare state that depended on the caring labour of women in the context of full male employment. Thomson also describes how, in the process, a new 'landscape of the child' was carved out, reconfiguring the parameters of the state, the home and urban space. Development of the young was to be fostered through a state-maintained framework of education, medicine, social services and economic policy, a nurturing infant-centred home and the provision of specially designed protective spaces (such as playgrounds and children's TV) away from risky adult environments. However, this new cradle of social citizenship was predicated on a so-called 'Golden Age' of industrial economy that, by the early 1970s, was lurching toward crisis. The manifest gender hierarchy and oppression underpinning embedded liberal ideals of the nuclear family also came under systematic attack from second-wave feminists, who exposed the darker side of family as a common site of abuse and violence against children.

By the 1980s, the New Right was railing against the impact of welfare benefits and the 'nanny state', invoking amongst other things the negative impact on children's moral development. Marking the activation of what Peck and Tickell (2002) characterise as first-phase 'roll back' neoliberalism, Keynesian models of social security were attacked as dysfunctional and encouraging a growing underclass dependent on state handouts. A proliferating 'rabble' of crime-prone sons and promiscuous daughters were predicted unless state handouts were diminished (Murray, 1994). Instead, it was argued, the family should be recognised as the bedrock of civilisation and left to fulfil its social responsibilities. The then Conservative prime minister Margaret Thatcher was among the more notable advocates of this view, as she articulated in a speech in 1988:

> The family is the building block of society. It's a nursery, a school, a hospital, a leisure place, a place of refuge and a place of rest. It encompasses the whole of society. It fashions beliefs. It's the preparation for the rest of our life and women run it.

But this traditional model of the dutiful family fitted neither the changing cultural attitudes of the day nor the transformative economic and social order Thatcher was to usher in. Deregulation, free marketeering, privatisation and a diminished state were promoted through a championing of ideals of freedom and liberation that cut across old expectations and obligations. As Nancy Fraser (2009) argues, feminist emancipatory ideals including critiques of 'the family' were appropriated and made mainstream through a centring of the self-determining, networked individual, liberated from gendered and classed expectations and ties.

Through the 1990s, the discrediting of Keynesian welfarism metamorphosed into phase-two 'roll out neoliberalism', encompassing a new logic of state interventionism (Peck and Tickell, 2002). By the time the New Labour government came to power in 1997, children were firmly positioned at the centre of a new neoliberal-inspired paradigm of social investment. Rather than supporting families to raise 'normal' minds and bodies, children became viewed as future assets which could be maximised for the good of all (Jenson, 2004). The state's role was no longer to act as an agent of social security but, instead, to enable personal responsibility and, crucially, to manage and prevent the social risks that might undermine children's future life chances. In the process, conceptualisations of family shifted away from the 'essential building block' metaphor toward a more contingent designation as, at one and the same time, a strength and a risk factor to be monitored and regulated.

Economic theorising around human capital was particularly influential in shifting the axis of concern away from a protective embedding of well-adjusted social citizenship toward the management of children as investment portfolios. Gary Becker (1981), for example, conceived of families as small factories, held together not by obligation or sentiment but by mutual interest in the human commodities they produce. Children came to assume a much greater significance within this market-based ethic as raw material requiring extensive investment to secure their futures as self-serving and self-producing subjects. And increasingly, social problems like poverty and inequality became framed in terms of lack of human capital, attributable to poor parenting. Amidst rising rates of divorce, cohabitation, birth outside of marriage and same-sex parenting, definitions of family became more flexible and inclusive – crucially, through a centring of child-rearing as the primary moral concern. The replacement of male breadwinner models of family with norms around dual-earner households was promoted as the progressive solution to gender injustice, while the female-dominated practice of childcare was redrawn as a motor of meritocracy.

Cognitive development was the core component of dominant conceptualisations of human capital at the beginning of the twenty-first century. Intensively parented children, it was argued, would be better able to navigate and capitalise on post-industrial opportunities. But the job of cultivating competent minds, fit to compete in the global knowledge economy was regarded as too important to be left to untrained parents. The New Labour years were characterised by a massive expansion of state-sponsored and third sector initiatives directly targeting families under the rubric of 'parenting support' (Edwards and Gillies, 2004). A new interventionist policy ethos began explicitly to position family life as a public rather than a private concern through the linking of parenting practices to broader narratives of social justice. As the minutiae of everyday relations with children came to be seen as directly determining their future outcomes, good parenting was made synonymous with maximising a child's cognitive potential and inculcating aspirational neoliberal values. Family households were rebranded as home learning environments, childcare became early education, and political consensus converged around the notion that parenting was the key to increasing social mobility.

Policy-makers were particularly impressed by the theorising of the US Nobel Laureate economist James Heckman. Arguing that human capital is cumulative rather than fixed, Heckman and colleagues proposed a formula summed up in the phrase 'skills beget skills and abilities beget abilities' (Cunha and Heckman, 2007). This economic reasoning, known as the 'Heckman equation', asserted that return on human capital was very high in the early years of life and diminished rapidly thereafter. In developing this model, Heckman and colleagues produced a graph that was to lever huge influence in social policy-making circles, inspiring New Labour's focus on early years provision as well as the subsequent coalition government's social mobility strategy and continued investment in preschool nursery provision (Wintour, 2012). Showing projected 'rates of return on investment in human capital by age', the image was widely reproduced as if it were proof in itself of the Heckman contentions (Howard-Jones *et al.*, 2012).

In 2006, the New Labour prime minister Tony Blair promoted his government's £17 billion investment in the Sure Start programme in a speech titled 'Our Nation's Future', citing the Heckman equation. Trailing a range of public service reforms, he concluded 'more than anything else, early intervention is crucial if we are to tackle social exclusion' (Blair, 2006). The subsequent policy focus on early childhood reflected the broader shift away from welfare state principles of shared responsibility and universal protection towards a preoccupation with identifying and managing individual risk factors (Featherstone *et al.*, 2014). More coercive policy approaches began to explicitly target disadvantaged mothers, positioning them as the essential mediators of their children's high-risk profile. Intensive family support initiatives were introduced with the promise of tackling recalcitrant parents and forcing them to parent more effectively (Gillies, 2011).

Beyond the economic theorising, there was little concrete evidence to support claims that altering parenting practices and maximising childhood investment could reverse structurally engrained inequality or even address educational attainment gaps. Longitudinal evaluations of interventions produced disappointing results (White and Wastell, 2015), while cohort studies continued to highlight the significance of income and maternal education above and beyond parenting styles (Hartas, 2011, 2012). But concern over this lack of verification from social research findings was overtaken by a new interest in psychologically inflected economic theorising associated with behavioural economics and, more specifically, models that grounded human behaviour in an emotional and social nexus (Jones *et al.*, 2013). Efforts to maximise the cognitive capacity of children were instead viewed through the lens of a more nuanced engagement with the psychological foundations of achievement.

Emotion regulation, resilience and empathetic connection became positioned as fundamental precursors to learning. Disadvantaged parents were accused not simply of insufficient cultivation of human capital, but of failing to equip children with a rational mindset capable of learning. Social and Emotional Aspects of Learning (SEAL) programmes were introduced in primary and secondary schools in an effort to address this perceived deficit, aiming to encourage personal control, motivation and 'empathy skills'. Simultaneously, the policy focus on parenting intensified through new conceptualisation of emotional impoverishment (McVarish *et al.*, 2015). A reanimated version of attachment theory readily explained this perceived emotional and intellectual impairment in terms of insensitive early years parenting (Thornton, 2011). More significantly, attachment as a process began to be articulated as an observably biological process, engraved into the structures of the developing brain.

Rescuing the infant brain

While parenting support was being developed as a key plank of New Labour's policy reforms in the 1990s, social investment spending in the US was coming under increasing attack. Evaluations of long-standing programmes like Head Start (the model for Sure Start) showed little impact on measurable outcomes for children, and many on the political right were dismissing the model as a waste of money. Advocates for childhood investment countered that interventions must begin at an earlier stage of development in order for them to be effective. Philanthropic organisations, such as the Carnegie Corporation and the Rob Reiner Foundation, made this argument by drawing on the language and imagery of neuroscience to suggest too much brain development had taken place by the age most interventions kicked in.

As John Bruer (1999) documents, this pseudoscientific explanation captured the public and political imagination, inspiring a 'common-sense' attribution of social problems to deficiencies in infant brain development. Neuro rhetoric dovetailed with a broader cultural fascination with the brain and was to drive a remarkably effective public relations campaign, attracting more wealthy philanthropists, charitable foundations and high-profile public figures. Another White House conference was convened in 1997, this time by the US President Bill Clinton and First Lady Hillary Clinton, to discuss early childhood development and the brain. In her opening address to the conference, Hillary Clinton stressed the importance of new insights into the contingent biological influence of the early years, noting that

> the song a father sings to his child in the morning, or a story that a mother reads to her child before bed help lay the foundation for a child's life, and in turn, for our nation's future. ... These experiences can determine whether children will grow up to be peaceful or violent

citizens, focused or undisciplined workers, attentive or detached parents themselves.

(cited in Bruer, 1999: 4–5)

Largely disconnected from a rapidly developing academic discipline of neuroscience, US child advocacy groups sprang up claiming to 'synthesise' and make accessible the latest biological research to highlight the unique potential and risks of the first three years of life. Sensitive mothering within this window of opportunity was promoted as making or breaking a child's future, 'hardwiring' them for success or failure.

By the mid 2000s, references to infant brain development were beginning to creep into UK policy documents, often lifted wholesale from the US advocacy groups (McVarish *et al.*, 2015). Such claims supplemented an already thriving policy consensus around the social and economic significance of parenting, yet their apparent grounding in science carried its own momentum, proving irresistibly appealing to a variety of UK-based philanthropists, politicians and public figures. Camila Batmanghelidjh, founder of the ill-fated charity Kids Company, was inspired to spend millions of the organisation's funds on brain scan research after receiving a 'sheaf of papers' from Prince Charles suggesting that childhood neglect changed brain structure (White, 2015). Her aim was to prove that the deprived young people who used Kids Company were psychologically damaged by their parents and requiring of specialist therapeutic help. She was vocal and passionate in her promotion of this theory, and lack of evidence from her scanning studies did little to dent her conviction.

Also captivated were the Conservative MP (and soon to be Work and Pensions Minister) Iain Duncan Smith and the Labour MP Graham Allen. Sharing Batmanghelidjh's belief in an organically damaged underclass, both politicians found accounts of infant brain science compelling. As the first rumblings of the global financial crisis were being felt, Duncan Smith and Allen were brought together by the WAVE Trust, a philanthrocapitalist organisation, to co-produce a paper attributing violence, low intelligence and poverty to the brain-stunting consequences of poor 'maternal attunement' (Duncan Smith and Allen, 2008). As instigators and ghost authors of this publication, the WAVE Trust began an unobtrusive but highly effective campaign to promote and spread this biologized cycle of deprivation narrative. Founded by business strategists with a self-proclaimed mission to apply the same approaches that turn loss-making companies profitable to social problems, they urged that 'root causes' of social disorder, rather than its symptoms, be tackled. The WAVE Trust campaigned within the higher echelons of power, reproducing the brain claims from the US child advocacy groups and embroidering them with eye-popping projections for return on investment. Ministers, civil servants, leading non-governmental organisations and like-minded philanthropists were lobbied to embrace a seductive scientific and financial logic promising that saving the brains of disadvantaged infants would slash costs for the public purse (Hosking and Walsh, 2005).

Austerity and early intervention

In the wake of the financial crisis and under the auspices of a new Conservative-led coalition government, economic theories became further psychologised and embedded within policy-making, reflecting mounting disillusionment with rational economic actor models (Davies, 2012). New attempts were made to envisage and requisition the social and emotional as underpinning 'judicious' choices that sustain the political equilibrium. More specifically, educational initiatives to embed emotional and social skills in the school curriculum morphed into a Victorian-flavoured preoccupation with the development of 'character' as a slippery inchoate vision of purposeful determination, self-direction and restraint, with a 'military ethos', most often projected on to white, public school boys (see Department for Education, 2014). At a broader level, public policy experimented with psychologised techniques of 'soft paternalism'; for example, through the commissioning of a Behavioural Insights Team, widely known as the 'nudge unit' (Jones *et al.*, 2013).

But in the realm of family policy, 'nudge' became shove as an ideological narrative of austerity drove brutal cuts to New Labour's social investment spending (Edwards and Gillies, 2016). Efforts to regulate the minds and brains of children became more assertive and explicitly targeted at disadvantaged mothers of children under two. As funding for children's centres and other universal services was slashed, creative appropriations of neuroscience were systematically worked into the coalition government's key child and family policy documents, justifying a narrowing down and intensification of intervention (e.g. see Allen, 2011; Field, 2010; Munro, 2011; Tickell, 2011). Envisaged in terms of an inoculation against irrationality and personal pathology, early intervention was firmly directed at those viewed as most likely to raise problem children.

A government-commission review on poverty and life chances concluded that 'the development of a baby's brain is affected by the attachment to their parents' and that brain growth is 'significantly reduced' in inadequately parented children (Field, 2010: 41). Similarly, the highly influential Allen review into early intervention called for urgent government action on the basis that 'brain architecture' is set during the first years, inside and outside the womb, with the 'wrong type' of parenting profoundly affecting children's 'emotional wiring' (Allen, 2011: xii). After the English riots of 2011 were diagnosed by the government as a 'crisis in parenting skills', a Troubled Families Unit was established with the aim of 'gripping' and 'turning around' parents identified as the wellspring of social disorder (mainly the sick, poor and disabled). In addition, the Family Nurse Partnership (FNP) programme, tasked with breaking intergenerational cycles of deprivation, was massively expanded. FNP practitioners identify 'at risk' pregnant women and visit them until the child's second birthday. Targeted mothers are trained to parent 'sensitively' for the sake of their children's neural development and cautioned about the brain-corroding effects of stress (Edwards *et al.*, 2015).

These and other simplistic misappropriations of neuroscience, rooted in the strategically developed claims of the US child advocacy groups, rapidly acquired the status of unchallenged fact in political, policy and practitioner circles. In 2013, the All-Party Parliamentary Group for Conception to Age Two was founded with support from all the major political organisations (co-chaired by Caroline Lucas from the Green Party). The WAVE Trust acted as the group secretariat, producing a cross-party manifesto that called for 'every baby to receive sensitive and responsive care from their main caregivers in the first years of life' (Leadsom *et al.*, 2013: 8). Their recommendations essentially amounted to greater monitoring of new mothers and intervention for those deemed insufficiently 'attuned'. Without these safeguards, the group warned, there would be 'another generation of disadvantage, inequality and dysfunction' (Howse, 2015). The manifesto included a foreword from Sally Davies, the Chief Medical Officer, decrying the 'cycle of harm' and declaring that 'science is helping us to understand how love and nurture by caring adults is hard wired into the brains of children' (Leadsom *et al.*, 2013: 2).

Infant brain determinism calcified into policy orthodoxy, impervious to the dubious provenance and misleading nature of the 'science' in question (McVarish *et al.*, 2015; Wastell and White, 2012; Edwards *et al.*, 2015; Bruer, 1999). Risk, as opposed to need, became the key justification for family services, with the defence of children's future prospects driving increasingly fervent and uncompromising forms of action. As Brid Featherstone and colleagues (2014) outline, an 'unholy alliance' formed between early intervention and child protection, shored up by neuroscientifically embellished narratives of 'now or never'. Chiming with broader austerity-inspired caricatures of the feckless poor, children from disadvantaged families increasingly became subject to a morally charged level of state surveillance. Concern that poor parents might wreak permanent damage on their children has since been reinforced through legislative changes criminalising an ill-defined category of 'emotional neglect' (Williams, 2014) and the introduction of timescales to speed up care proceedings (Featherstone, 2014). Following these changes and several high-profile child abuse cases, a steep incline began in the numbers of children taken from their families and placed into state care, with official statistics reaching new records year on year (Department for Education, 2015).

Meanwhile, government ministers have pursued a strategy designed to accelerate and escalate the numbers of children made available for adoption. New guidance and funding promoting the swift removal of 'at risk' children and their resettlement with a new family was issued to local authorities in 2011. Michael Gove, then Conservative Education Minister in the coalition government and responsible for crafting legislative reforms on the issue, emphasised the need for 'social workers to feel empowered to use robust measures with those parents who won't shape up' (2012). Hailing the transformative powers of adoption, he pledged to address the 'cruel rationing of human love for those most in need' (Gove, 2012). A new 'foster to adopt clause' was enshrined in the 2014 Children and Families Act, requiring looked-after children to be placed with prospective new families before the onset of legal procedures.

Rates of adoption initially rose to record levels, only to fall back sharply when the Court of Appeal issued a stern judgement rebuking inattention to human rights (Family Law Week, 2013). As Featherstone and Bywaters (2014) point out, an ideological commitment to adoption is being pursued alongside unprecedented cuts to family support that have left many parents without adequate resources to care for their children. Particular criticism has been directed at the UK from the European Council for removing children from mothers who had experienced domestic violence and depression. Nevertheless, the subsequent dip in numbers was widely described as a 'crisis' by the media and as a 'tragedy' by Prime Minister David Cameron (2015), who promised new measures in the forthcoming Education and Adoption Bill that would allow government to intervene directly to speed up local authority adoption services. In the context of austerity, adoption is promoted as financially prudent, morally right and transformative for the children concerned (Narey, 2011). Alternative, humane models of child protection addressing the economic, material and social support needs of struggling families have been increasingly marginalised in the context of 'shape up or ship out' ultimatums (Featherstone *et al.*, 2014).

Back to the future? From risk to resilience

As the neoliberal paradigm is buffeted by global economic crises, children have been increasingly targeted as a core resource through which market-based rationality can be anchored. This has played out through a revival of Victorian themes about the importance of nurturing personal traits of determination and resilience alongside a renewed mission to rescue poor children from their irredeemable parents. This represents more than a harking back to another age. Possessive individualism has been reworked to reflect a different political agenda as well as more contemporary sensibilities and concerns around the psychological development of the young, the prevention of permanent harm, and the responsibility of parents and the state. The positioning of children as national capital has rendered them public property, justifying the policing of parents in the name of early intervention. As such, the state is mobilised on behalf of the market to secure the production of clear-thinking, flexible, self-directed brains able to withstand the pressures of a global competitive system.

More significantly, advanced neoliberalism has become entirely detached from the classical liberal belief that market-based logic is rooted in human nature and realisable only when free from the distortions of the state (Soss *et al.*, 2011). Instead, market behaviour is perceived as learnt rather than natural, requiring the firm hand of government to secure the future though childhood intervention. Evaluations of personhood that were once essentially moral have since been psychologised and re-presented in terms of emotional and cognitive capabilities. Traditional understandings of 'character' conveyed strong notions of moral virtue, whereas contemporary invocations denote personal competence and wellbeing. Meanwhile, the process of psychological development itself has become deeply moralised as an aspirational goal rather than an end in itself. As Kathryn Ecclestone (2012) notes, the imperative has shifted to the process of acquiring traits defined as character, with the onus placed on the development of appropriate capabilities. Morality, then, is no longer assumed to inhere in any trait, capability or personality but, rather, in the very act of transferring value to the self. Laissez-faire optimism that poverty and destitution would be eradicated through a strengthening of moral fibre has been replaced by a more pessimistic drive to equip developing minds and brains with the psychological tools to endure uncertainty, hardship and distress (see also Ecclestone, this volume).

This reconfiguration and redeployment of Victorian morality in relation to children has followed through into old-style practices of child rescue. Tussles are now over brains rather than souls, but the impassioned justifications for action remain remarkably similar in tone if not content. Legislative efforts to increase adoption have been widely described as a 'crusade', with a 'loving home' offered by a white, middle-class family uncritically presented as best for the child (Barn, 2013). Leading the call for greater numbers of children to be taken from their unsatisfactory parents was Martin Narey, a Chief Executive of Barnardo's until he retired to become the government's 'adoption tsar' in 2011. Unlike his forebear Thomas Barnardo, Narey's mission is pre-emptive rather than rehabilitative, aimed at increasing the removal of babies at birth to prevent them being damaged beyond repair by inadequate parents. This reflects the particular significance accorded to time and risk within asset building rationales of human capital. Concern is projected onto what children will become in the future rather than what they are experiencing in the here and now. While Victorian child-centric reformers expressed moral repugnance at the suffering of vulnerable children, contemporary child savers denounce the negative effects deprivation will have on their later life chances. Narey (2015) is explicit about this in his 'blueprint' for adoption reform:

> I have intentionally talked about productivity even though I'm aware that many practitioners object to the application of such a term to an issue as sensitive as a child's future. But this is very much about productivity because delay is so damaging to children.

As this chapter demonstrates, children's development has been targeted as raw potential since the late nineteenth century with varying political and economic models driving frameworks of intervention. Consistent across time has been the notion that family relationships can be re-engineered, optimised or replaced to tackle social and structural problems. Wealthy philanthropists and social reformers in particular have been instrumental in championing simple solutions that promise to breed out poverty and crime at the level of the family without recourse to redistributive solutions. Theorising around degenerate character, physical weakness, genetic inferiority and psychological maladjustment represents efforts to tackle the shortcomings of capitalism by nurturing stronger, more resilient subjects. The contemporary policy preoccupation with suboptimal infant brains is merely the latest incarnation of a long-standing conviction held by the rich and powerful – specifically, that there must be something inherently wrong with the minds, bodies and souls of those failing to thrive in an unfettered free market economy.

References

Allen, G. (2011) Early Intervention: The Next Steps. An Independent Report to Her Majesty's Government. Cabinet Office, London.

Barn, R. (2013) '"Race" and Adoption Crusades in 21st Century Britain', *Huffington Post*, Updated 14 April. Available at: www.huffingtonpost.co.uk/professor-ravin der-barn/race-and-adoption-michael-gove_b_2657383.html

Barnardo, T. (1885) *Worse Than Orphans: How I Stole Two Girls and Fought for a Boy*. J. F. Shaw & Company, London.

Becker, G. (1981) *Treatise on the Family*. Harvard University Press, Cambridge, MA.

Behlmer, G. K. (1998) *Friends of the Family: The English Home and its Guardians, 1850–1940*. Stanford University Press, Stanford, CA.

Blair, T. (2006) Our nation's future – social exclusion, Speech, York, 5 September. Available at: http://webarchive.nationalarchives.gov.uk/20040105034004/http:/num ber10.gov.uk/page10037

Boucher, E. (2014) *Empire's Children: Child Emigration, Welfare, and the Decline of the British World, 1869–1967*, Cambridge University Press, Cambridge.

Bruer, J. T. (1999) *The Myth of the First Three Years: A New Understanding of Early Brain Development and Lifelong Learning*. Simon & Schuster, New York.

Cunha, F. and Heckman, J. (2007) 'The technology of skill formation'. *American Economic Review*, 97(2): 31–47.

Davies, W. (2012) 'The Emerging Neocommunitarianism'. *Political Quarterly*, 83(4): 767–776.

Department for Education (2014) 'Measures to help schools instil character in pupils announced'. Press Release, 8 December. Available at: https://www.gov.uk/governm ent/news/measures-to-help-schools-instil-character-in-pupils-announced

Department for Education (2015) Children looked after in England (including adoption and care leavers) year ending 31 March 2015. SFR 34/2015. Available at: https://www. gov.uk/government/uploads/system/uploads/attachment_data/file/464756/SFR34_ 2015_Text.pdf

Dikötter, F. (1998) 'Race culture: recent perspectives on the history of eugenics'. *The American Historical Review*, 103(2): 467–478.

Duncan Smith, I. and Allen, G. (2008) Early Intervention: Good Parents, Great Kids, Better Citizens. Centre for Social Justice and Smith Institute, London.

Ecclestone, K. (2012) 'From emotional and psychological well-being to character education: challenging policy discourses of behavioural science and "vulnerability"'. *Research Papers in Education*, 27(4): 463–480.

Edwards, R. and Gillies, V. (2004) 'Support in parenting: values and consensus concerning who to turn to'. *Journal of Social Policy*, 33(4): 623–643.

Edwards, R. and Gillies, V. (2016) 'Family policy: the Mods and Rockers', in H. Bochel and M. Powell (Eds.) *The Coalition Government and Social Policy.* Policy Press, Bristol, pp. 243–264.

Edwards, R., Gillies, V. and Horsley, N. (2015) 'Brain science and early years policy: hopeful ethos or "cruel optimism"?' *Critical Social Policy*, 35(2): 167–187.

Evans, B., Rahman, S. and Jones, E. (2008) 'Managing the "unmanageable": interwar child psychiatry at the Maudsley Hospital'. *History of Psychiatry*, 19(4): 454–475.

Family Law Week (2013) 'Court of Appeal gives important guidance on adoption applications' [online]. *Family Law Week*, 22 September. Available at: www.familyla wweek.co.uk/site.aspx?i=ed117222

Featherstone, B. and Bywaters, P. (2014) 'An ideological approach to adoption figures means we are missing important trends'. *Community Care*, 7 October. Available at: www.communitycare.co.uk/2014/10/07/ideological-approach-adoption-figures-means-missing-important-trends/

Featherstone, B., Morris, K. and White, S. (2014) 'A marriage made in hell: early intervention meets child protection'. *British Journal of Social Work*, 44(7): 1735–1749.

Field, F. (2010) The Foundation Years: Preventing Poor Children Becoming Poor Adults. The Report of the Independent Review on Poverty and Life Chances. Cabinet Office, London.

Fraser, N. (2009) 'Feminism, capitalism and the cunning of history'. *New Left Review*, 56: 97–117.

Gillies, V. (2011) 'From function to competence: engaging with the new politics of family' [online]. *Sociological Research Online*, 16(4). Available at: www.socresonline. org.uk/16/4/11.html

Gove, M. (2012) Speech on adoption, given at the Isaac Newton Centre for Continuing Professional Development, 23 February. Available at: https://www.gov.uk/gov ernment/speeches/michael-gove-speech-on-adoption.

Guardian, The (2015) 'David Cameron urges faster adoptions doubling number of early placements', *The Guardian*, 2 November. Available at: www.theguardian.com/ society/2015/nov/02/david-cameron-urges-faster-adoptions-doubling-number-of-early-placements?CMP=share_btn_tw

Hartas, D. (2011) 'Families' social backgrounds matter: socio-economic factors, home learning and young children's language, literacy and social outcomes'. *British Educational Research Journal*, 37(6): 893–914.

Hartas, D. (2012) 'Inequality and the home learning environment: predictions about seven-year olds' language and literacy'. *British Educational Research Journal*, 38(5): 859–879.

Hosking, G. and Walsh, I. (2005) 'The WAVE Report 2005: Violence and What to Do About It', WAVE Trust, Croydon. Available at: www.wavetrust.org/sites/default/ files/reports/migrate-wave-report-2005-full-report.pdf

Howard-Jones, P. A., Washbrook, E. V. and Meadows, S. (2012) 'The timing of educational investment: a neuroscientific perspective'. *Developmental Cognitive Neuroscience*, 2(1): 18–29.

Howse, P. (2015) 'Invest in the first 1001 days say experts' [online]. *BBC News*, 25 February. Available at: www.bbc.co.uk/news/education-31607711

Jenson, J. (2004) 'Changing the paradigm: family responsibility or investing in children'. *The Canadian Journal of Sociology*, 29(2): 169–192.

Jenson, T. and Tyler, I. (2015) 'Benefits broods: the cultural and political crafting of anti-welfare commonsense'. *Critical Social Policy*, 35(4): 1–22.

Jones, R., Pykett, J. and Whitehead, M. (2013) *Changing Behaviours: On the Rise of the Psychological State.* Edward Elgar, Cheltenham.

Kagan, J. (1998) *Three Seductive Ideas.* Harvard University Press, Cambridge, MA.

Kent, K. S. (2002) *Gender and Power in Britain 1640–1990.* Routledge, London.

Kuchta, T. (2010) *Semi-detached Empire: Suburbia and the Colonization of Britain, 1880 to the Present.* Virginia University Press, Charlottesville, VA.

Leadsom, A., Field, F., Burstow, P. and Lucas, C. (2013) The 1001 Critical Days. The Importance of the Conception to Age Two Period. A Cross-party Manifesto [online]. Available at: www.1001criticaldays.co.uk/1001days_Nov15.pdf

Liverpool Maritime Archive and Library (2014) Information Sheet 10: Child Emigration [online]. Available at: www.liverpoolmuseums.org.uk/maritime/archive/sheet/10

McVarish, J., Lee, E. and Lowe, P. (2015) 'Neuroscience and family policy: what becomes of the parent?' *Critical Social Policy*, 35(2): 248–269.

Munro, E. (2011) Munro Review of Child Protection: Final Report – A Child-centred System, Department for Education, London.

Murdoch, L. (2006) *Imagined Orphans: Poor Families, Child Welfare, and Contested Citizenship in London.* Rutgers, New Brunswick.

Murray, C. (1994) Underclass: The Crisis Deepens. Institute of Economic Affairs, London.

Narey, M. (2011) 'The Narey Report: a blueprint for the nation's lost children'. *The Times*, 1 November. Available at: www.thetimes.co.uk/tto/life/families/article308 3832.ece

Peck, J. and Tickell, A. (2002) 'Neoliberalizing space'. *Antipode*, 34(3): 380–404.

Rooff, M. (1972) *A Hundred Years of Family Welfare.* Michael Joseph, London.

Rose, N. (1999) *Governing the Soul: The Shaping of the Private Self.* Routledge, London.

Rosenberg, E. (2002) 'Missions to the World: Philanthropy Abroad', in L. J. Freedmand and M. D. McGarvie (Eds.) *Charity, Philanthropy, and Civility in American History.* Cambridge University Press, Cambridge, pp. 241–258.

Shapira, M. (2015) *The War Inside Psychoanalysis, Total War, and the Making of the Democratic Self in Postwar Britain.* Cambridge University Press, Cambridge.

Smuts, A. (2008) *Science in the Service of Children, 1893–1935.* Yale University Press, London.

Soss, J., Fording, R. and Schram, S. (2011) *Disciplining the Poor: Neoliberal Paternalism and the Persistent Power of Race.* University of Chicago Press, Chicago.

Stedman-Jones, G. (2012) *Outcast London: A Study in the Relationship between Classes in Victorian Society.* Verso, London.

Stewart, J. (2013) *Child Guidance in Britain, 1918–1955: The Dangerous Age of Childhood.* Routledge, Abingdon.

Swain, S. (2011) 'Failing families: echoes of nineteenth century discourse in contemporary debates around child protection', in M. L. Lohlke and C. Gutleben (Eds.) *Neo-Victorian Families: Gender, Sexual and Cultural Politics.* Rodopi, Amsterdam, pp. 71–92.

Thatcher, M. (1988) Speech to the Conservative Women's Conference, Barbican Centre, London, 25 May. Available at: www.margaretthatcher.org/document/107248

Thomson, M. (2013) *Lost Freedom: The Landscape of the Child and the British Post-War Settlement.* Oxford University Press, Oxford.

Thornton, D. J. (2011) 'Neuroscience, affect and the entrepreneurialisation of motherhood'. *Communication and Critical/Cultural Studies* 8(4): 399–424.

Tickell, DameC. (2011) The Early Years Foundation Stage: Review Report on the Evidence. Department for Education, London.

Wastell, D. and White, S. (2012) 'Blinded by neuroscience: social policy, the family and the infant brain'. *Families, Relationships and Societies*, 1(3): 397–415.

White, A. (2015) 'Questions raised over Kids Company spending and research', *Buzzfeed News*, 6 July. Available at: www.buzzfeed.com/alanwhite/questions-raise d-over-kids-company-spending-and-research#.qu3P90Kvv

White, S. and Wastell, D. (2015) 'The rise and rise of prevention science in UK family welfare: surveillance gets under the skin' [online]. *Families, Relationships and Societies,* 24 November. Available at: http://dx.doi.org/10.1332/204674315X14479283041843

Williams, M. (2014) 'Child neglect: a change in the law' [online]. *Society Central*, 31 March. Available at: https://societycentral.ac.uk/2014/03/31/child-neglect-a-change-in-the-law/

Wintour, P. (2012) 'Apparent consensus on social mobility masks fundamental split', *The Guardian*, 29 May. Available at: www.theguardian.com/society/2012/may/29/consensus-social-mobility-fundamental-split

7 The imperative to shape young brains
Mindfulness as a neuroeducational intervention

Alberto Sánchez-Allred and Suparna Choudhury

Introduction: neuroscience and the "therapeutic turn" in education

Developmental brain science has recently come together with Buddhist-derived meditation, leading to a less than obvious new candidate for educational programs for children and adolescents: brain-based mindfulness techniques. In this chapter, we explore the scientific and cultural contexts that have made possible this emerging trend in education and propose reasons for the enormous excitement and investment surrounding it. Teaching children and adolescents mindfulness meditation in school is a reflection of recent dialogue between neuroscientists and mindfulness practitioners and promoters as well as of a desire for new pedagogical tools, cast in neurological terms, that address the moral and emotional formation of children. Although from a strictly scientific perspective, it is premature to state that mindfulness meditation accomplishes the educational goals toward which it is being implemented in schools across North America and the UK, it is nonetheless being used as a practise that aims to shape "better" (i.e. more resilient, emotionally intelligent, and well-regulated) modern humans as they develop. While current research projects such as The Oxford Mindfulness and Resilience in Adolescence study seek to answer the effectiveness question, we are more concerned with addressing the issue of how it is that mindfulness has come to be promoted in education at this particular moment in time.

The trend of adopting mindfulness in the classroom may be understood through the lens of what some researchers have identified as a "therapeutic turn" in education (Ecclestone and Hayes, 2009). It is indicative of a wider therapeutic behaviour change agenda closely associated with the notion of psychological governance developed in this book (see Ecclestone, this volume). This describes a host of new pedagogical interventions that prioritize the emotional well-being of children. The backdrop evoked by such interventions is one of crisis, with the educational system not adequately responding to pedagogical needs of children who are viewed as inadequately equipped to handle the demands of modern life. Mindfulness, as a psychological and emotional intervention, is promoted to render the educational system more effective in achieving both traditional and more contemporary goals; according to its proponents, "mindful" teachers and students may be better able to succeed not only in reading and arithmetic but also in building resilience, through the acquisition of tools for weathering the stresses of modern life, and in developing effective emotional self-regulation, particularly as it is viewed in the context of a proper relationship to others and oneself. That last goal, of course, is traditionally the domain of morality and ethics, and it is for this reason that mindfulness is also being promoted under the rubric of moral and ethical education (Roeser *et al.*, 2014).

How is it, then, that a secular and science-based version of mindfulness has entered the realm of contemporary education? The convergence of Buddhism and neuroscience through the flourishing field of consciousness studies during recent decades provides a premise to this trend, making the link between the brain and meditation almost self-evident. Several research groups in laboratories around the world have focused their investigations on the "meditating brain" using neuroimaging techniques, with questions about empathy, attention networks, the brain's "default mode," the self, and mind-wandering driving their questions. Furthermore, the increasing trend to take embodied and extended mind approaches in neuroscience converges with, and is even inspired by, theory from Buddhism (these connections are developed explicitly within education by Claxton, 1998). Simultaneously, education researchers have come together with neuroscientists seeking scientific guidance for effective pedagogical strategies based on the latest brain research. Of course, there is considerable "neuro-hype," motivating – and at times distorting – neuroeducation and educational brain-based strategies and policies. Even in the case of mindfulness, where there is a body of research that suggests significant and positive effects on the brain, the science is not yet in a position to adequately inform interventions. Nonetheless, we illustrate how the "plastic" brain provides the conceptual basis by which educators and even children and adolescents themselves can imagine the effects of mindfulness as an intervention. Neuroeducation, especially in putting forward mindfulness as a scientifically valid neurological intervention, is in this respect a crucial support for the ambitions of the "therapeutic turn" and for modern moral and ethical education.

Although others have pointed toward a larger and more general shift toward "therapeutic societies" as a way of explaining how mindfulness and other practices and attitudes have come to be adopted in education (Ecclestone and Hayes, 2009; Furedi, 2003; Nolan, 1998), we identify a discourse about children and adolescents, informed by findings related to neuroplasticity, that creates an imperative for practices that can intervene and promise positive outcomes for this demographic. As such, the example of mindfulness interventions as a tool of neuroeducation can be analysed under the broader rubric of psychological governance as a strategy for molding the adolescent brain, behaviour, character and resilience. It has been suggested that teaching "critical thinking skills" has reached a dead end, that as a strategy, it was largely ineffective in achieving the desired goals (namely actions based on greater reflection), and that a move to virtue ethics within philosophy and the teaching of morality and ethical practices in schools, as it relates to "character" develop-ment, could provide a more effective way forward (Haidt, 2006). Mindfulness, in this regard, provides an excellent test case. Mindfulness is promoted as an evidence-based practice to shape brains and, by extension, the "character" of children, conceived in terms of emotional regulation. Our research indicates that teachers and students think about mindfulness according to its presumed impacts on the brain and the consequences of these impacts in terms of resi-lience and the "executive" control and modulation of emotions and actions. However, it is only after decades of heavy investments in neuroscientific research, along with the popularization of the "plastic" brain and mind-fulness as a brain-transforming practice, that it is possible to have mindfulness promoted as a scientific pedagogical intervention for the ethical development and emotional well-being of schoolchildren. Notions of "working on the self" by means of "sculpting" or "shaping the brain" with measurable outcomes are made possible through the conceptual anchor of plasticity (Pitts-Taylor, 2010); training young people in scientifically informed mindfulness programs provides the possibility of shaping young brains with the particular goals of cultivating self-regulation, emotional intelligence and resilience.

In this chapter, we present insights from an ongoing study of the social and cultural contexts of neuroeducation involving teachers and mentors working with young people using mindfulness education. By examining trends in neu-roscience, education and policy, we propose reasons that explain the recent excitement around mindfulness-based education for children and adolescents. In particular, we suggest that the neuroplastic understanding of the brain provides state-of-the-art evidence for practices capable of molding young brains, fulfilling long-standing concerns about the social, moral and cognitive development of children and adolescents. We suggest that what is actually molded, through mindfulness, is the brain as an ethical substance. We discuss how the widespread appeal of mindfulness as a brain-based technique is symptomatic of "eastward journeys" – the melding of Eastern philosophy and medicine – to remedy existential deficiencies in the West. Furthermore, the convergence of spirituality, malleable brains and ethical development of young people provides powerful mechanistic notions of hope and possibility for the futures of children and adolescence as well as for education.

The rising influence of the brain sciences in health and society

A rising interest in "neuropolicy" in North America and the UK during the last two decades has generated increasing imperatives among researchers to address the potential of brain research for policy-relevant applications. This has occurred amongst technological developments in neuroscience during the last 20 years that have generated enormous excitement about the potential of neuroscientific insights for mental health, childrearing, law, education and not least for revolutionizing existing visions of human nature (Changeaux, 1985; Bush, 1990; Wolpe, 2002). Large-scale initiatives such as the US-led Brain

Initiative and the European Human Brain Project (Markram, 2012) have been accompanied by major drives to increase knowledge translation, or public engagement, in order to facilitate dialogue between new "neuroscience communication experts" and the public about the goals, applications, and ethics of emerging neuroscience research (Illes *et al.*, 2010). This rapid rise in neuroscience research in laboratories across the globe along with the growing public awareness of neuroscientific knowledge in, at least, industrialized Western countries have ushered in a new wave of neuroscientific expertise led by scientists whose interpretations about brain data can provide insight into how to conduct one's life, frame the understanding and treatment of mental disorders, and furnish evidence with which to guide policy-making (Racine *et al.*, 2010).

In the literature on health policy and social studies of medicine, there has been much recent interest in how the rapid growth of neuroscience research and knowledge is reshaping our views not only of mental health and illness but also of human development and subjectivity more broadly. The findings of neuroscience and the widespread use of treatments and interventions based on brain research have given rise to a popular genre of "neurotalk," which is invoked to make sense of the range of human problems and predicaments (Racine *et al.*, 2010). Both neurotechnology and neurotalk are changing the ways we think of ourselves, our problems and our prospects. Interpretations of neuroimaging data are in this way influencing many aspects of everyday life including notions of maturity and normality, explanations for behaviour, modes of self-regulation and learning, all of which have recently been applied to several sites of public life pertaining to young people, including the psychiatric clinic, the law, parenting and education. Adolescence is a case example of an area of increasing interest for the transfer of brain-based knowledge to the development of policies, legislature, everyday practices and lay understandings concerning the management of behaviour.

Translating brain data for clinical and public policy

The promissory tones that have fuelled cognitive neuroscience to date are simultaneously tempered by cautionary (e.g. Farah, 2005), cynical (e.g. Tallis, 2011) and self-reflective (e.g. Choudhury and Slaby, 2012) voices, emphasizing the imbalance between the hopes for and outcomes of the enormous investment in the neuroscience of behaviour, which has often unfolded at the expense of alternative approaches. In particular, the relevance of neuroscientific findings for public policy ventures and clinical practices that affect individuals on a day-to-day basis (Ortega and Vidal, 2011) – such as how we should treat mental disorders, direct education and determine legal responsibility – have been challenged.

In spite of the preliminary nature of data from neuroscience and the cautioning of some neuroscientists about the swift application of this data, adolescent brain research is characteristic of a growing relationship between brain science and policy. Emerging critiques and commentaries from the social studies of neuroscience (Pickersgill *et al.*, 2015; Rose and Abi-Rached, 2013) and from within neuroscience (Choudhury, 2010; Raz, 2012; Weisberg *et al.*, 2008) have begun to highlight the limits of predominant methodologies as well as the ideological sea changes behind the rise of the new neuro-scientism, giving us reason to be cautious about the rapid entry of neuroscientific wisdom into social practices, particularly against a background of a strong media appetite for neuro-essentialism (Racine *et al.*, 2010). Studies have shown that this growing "neuro-hype" is tied to the "seductive allure" of neuroscientific explanations (Weisberg *et al.*, 2008) and the persuasiveness of brain images in rendering theories about behavior more objective and thus more credible (McCabe and Castel, 2008). A recent survey of scientific literature dealing with attention deficit/hyperactivity disorder (ADHD) demonstrated that it is neither simply the lay appetite for "hard" brain facts nor the media-driven distortions of neuroscience experiments that are solely responsible for dis-proportionate neuro-hype. The authors showed that data misrepresentations, extrapolations and over-promotion of preliminary data occur within scholarly research articles as well, contributing to the distorted appropriation of brain data – as unsubstantiated "neuromyths" – in wider circles (Gonon *et al.*, 2011).

Such studies caution scientists and the public that, amidst the push for public awareness of neuroscience and the pull for policy applications, knowledge translation can have deleterious consequences (e.g. Bruer, 1997; Knudsen, 2004). Our objective here is not so much to consider the shortcomings of neuroscientific evidence for educational curricula but, rather, to use a critical neuroscience (Choudhury and Slaby, 2012) approach to propose reasons for the mutual embrace between brain science and education in contemporary schools in the UK and North America. In particular, we focus on a growing interest in incorporating mindfulness-based education into evidence-based pedagogy.

The institution of neuroeducation

The field of neuroeducation has emerged in response to the drive for translational neuroscience in conjunction with a trend towards evidence-based education. It provides a rich site to explore how developmental brain data are transferred towards practical goals geared towards young people. Increasingly institutio-nalized through highly funded multidisciplinary centres, programs, academic journals and books, neuroeducation brings together researchers and educators to try to create new pathways between scientific research and educational practice based on a rigorous "learning science" (Battro *et al.*, 2008; Goswami, 2006). A key goal of this research agenda is to build effective strategies of education that improve the academic achievement and social-emotional development of young people, reframing curricula in terms of research on the developing brain. These strategies are presented as being based on science and make strong claims about what can be accomplished through programs that capitalize on brain plasticity. Although the science is relatively new, recently established high-profile graduate centres for neuroeducation in the

US, research centres in the UK, and burgeoning schools and programs for children in Canada have been set up to develop skills and capacities to bridge neuroscience and educational policy and to respond to varying degrees of concern that education is in crisis. The primary sites of the application of neuroscience research to education include reading/literacy, numeracy/arithmetic and "brain training" (Ansari *et al.*, 2011). Among some of the most promising recent applications of neuroscience to education are the non-symbolic representation of quantity (Cantlon *et al.*, 2006) and the use of this research with kindergartners (Wilson *et al.*, 2009), studies on exercise's impact on learning (Hillman *et al.*, 2008), and classroom interventions into developmental dyslexia based on brain imaging (Shaywitz and Shaywitz, 2004; Sarkari *et al.*, 2002; Temple *et al.*, 2003).

Despite the preliminary nature of research data, findings from neuroeducation have begun to trickle directly into schools in the form of new customized teaching methods built around interpretations of data about social cognitive development and neurocognitive plasticity. These methods aim predominantly to improve the ability of children and adolescents to control actions, impulses, and emotions and to provide a biological foundation for working with developmental disorders such as dyslexia in the classroom (Howard-Jones, 2009; Thomson *et al.*, 2013). This expeditious translation is not without controversy. Earlier attempts to bridge brain science and education through high-profile commercial projects, such as Baby Einstein and Baby Mozart, and educational programs premised on left brain/right brain lateralization have been widely discredited (Fischer, 2009; Maxwell and Racine, 2012), raising the need for greater scrutiny in the transfer of research to policy and educational practice. Critics highlight the commercialization and oversimplification of neuroscientific results and the emergence of "neuromyths," a term that gained prevalence with the Brain and Learning project report by the OECD in 2002 that refers specifically to "a misconception generated by a misunderstanding, a misreading, or a misquoting of facts scientifically established (by brain research) to make a case for use of brain research, in education and other contexts" (2002: 111). While researchers caution policy-makers about the "seductive allure" (Weisberg *et al.*, 2008) of "neuromythology" (Rose, 2005; Tallis, 2004) and the risks of applying preliminary findings too soon, the values and motives underlying the imperative for knowledge translation between researchers, policy-makers and teachers remain unexplored.

Mindfulness meditation as brain-based pedagogy

The field of mindfulness-based education has grown considerably in the last few years, progressing from being viewed as a "new age" marginal phenomenon on the fringes of educational practices to being increasingly accepted by the mainstream with funding agencies recently investing significantly in randomized controlled trials to analyze its long-term benefits on developing children and adolescents. Mindfulness-based education has become an institution, generating texts written by scientists, workshops on neuroscience and mindfulness for educators, new educational curricula and networks in the US, Canada and the UK, and non-profit organizations in the US that contend with poverty and behavioral problems among youth by using mindfulness and yoga. In the last few years, new curricula, new courses and even new schools have been explicitly established with the goal of equipping children and teenagers with the ability to develop and practice mindfulness as a tool for reducing stress, improving focus, regulating emotions and building resiliency. Most recently, in 2015, the Wellcome Trust has invested £6.4 million in a seven-year randomized controlled trial (RCT) run by British universities, which will be the first large-scale investigation into the effects of mindfulness-based education as an intervention for adolescents at the level of brain, behavior and cognition (Wellcome Trust, 2015). The objective is to understand to what extent, by what mechanisms and in which time period mindfulness techniques can improve self-regulation, emotional resilience and symptoms of depression and anxiety.

In the US, the Blue School, a "progressive independent school" in lower Manhattan, focuses as much attention on nurturing compassion and human relationships as it does on learning to read and write. The founders describe the establishment of the school as a response to the unsustainable and disharmonious nature of the world, against which children will need to be equipped with the skills of creativity, innovation and emotional resilience. Similarly, the .b (pronounced dot-be) programme (which stands for Stop, Breathe and Be!) – a nine-week course crafted by teachers in the UK in 2007 and introduced in schools across the UK as well as in other European countries – helps children and adolescents cope with the stresses of school life such as those induced by tests and bullying; improve their social interaction with teachers and peers; enhance focus and attention; and achieve greater happiness, calm, and fulfilment. The "7/11" and "beditation" techniques are becoming familiar exercises for about 3,000 students in Britain who have been introduced to mindfulness teachings. Similarly, "the breathing song" and "elevator breath" are phrases pupils in California are starting to use in schools. Mindfulness in Education teachers report using bells, reading stories and offering relaxation practices, yoga, centering and breath focus in their classes.

Adolescent neuroplasticity: teen brains "under construction"

Many of these programs emphasize the benefits of mindfulness-based activities for the still-developing adolescent brain. For example, the developers of Learning2Breathe, a mindfulness curriculum for adolescents, describe the program as being based on research, citing evidence that "adolescent brains are still under construction" and that teaching mindfulness can help cultivate skills for "emotion management" as neural pathways are developing (see Learning2Breathe, 2016). To date, most of the scientific literature on the subject cites associations between areas of the brain that are still developing in adolescence and those that are activated through mindfulness.

Researchers at Harvard and Cambridge are currently conducting RCTs to attempt to evaluate the causal effects of mindfulness on the brain. In a recently published white paper on integrating mindfulness into K-12 education in the US, which reviews the links between research on the developing brain, stress physiology and the effects of mindfulness training, the connection between mindfulness practices and the developing brain are also addressed; the paper arrives at the conclusion that "the brain regions that are impacted by mindfulness training are implicated in executive functioning (EF) and the regulation of emotions and behaviour" (Meiklejohn *et al.*, 2012: 293).

The literature and findings outlined above generally posit adolescent brains as "brains in construction," based on a new and active field of research involving neuroimaging that has revealed the considerable extent to which the brain matures in terms of structure and function in late childhood and throughout adolescence (Giedd and Denker, 2015). The studies repeatedly demonstrate that areas of the brain related to social cognition and executive function – primarily prefrontal cortex – are particularly implicated in this developmental growth (Blakemore and Choudhury, 2004). New neuroimaging (MRI and fMRI) data have shown that middle childhood to late adolescence represents a "sensitive period" of malleability – or neuroplasticity – during which environmental input can have particularly profound effects (Blakemore and Frith, 2005). Scientifically structured programs with mindfulness-centered curricula cite this impressionability as a rationale for their teaching strategies. Moreover, communicating to young people the idea of caring for, and shaping, one's own brain has become a new imperative in scientific outreach programs (Choudhury *et al.*, 2012). It has in this way become common for educators in such programs to be literate in neuroscience terms and for the programs themselves to explicitly define their pedagogical approaches in terms of neurological science. Children at the Blue School, for example, are taught impulse control, empathy and emotional regulation in a curriculum that draws heavily on research from cognitive neuroscience (the school's board members include David Rock, CEO of the Neuroleadership Group, and Dan Siegel, Clinical Professor of Psychiatry at UCLA). Other programs go a step further – not only are their pedagogical strategies informed by brain development research but also the students are taught neurological vocabulary as part of the educational process. Teachers trained to teach curricula such as MindUP are required to educate children about brain regions and their functions. MindUP prescribes the employment of a hand model (i.e. using the fingers and thumb) to visually demonstrate how frontal regulation of limbic areas, or what it designates the "upstairs brain" and the "downstairs brain," can contribute to the management of social, emotional and moral behaviors. A facility with the language of the brain is central to intervention goals.

Turning to philosophy, both West and East

While mindfulness programs in education are promoted as objective and scientific, they include implicit moral assumptions about the kinds of young people we ought to cultivate. Researchers and policy-makers repeatedly refer to emotional regulation as a core value of mindfulness projects, emphasizing that emotional development is a necessary, progressive and new imperative of contemporary science-based educational programs (Hyland, 2011). In the language of one program, the Mindfulness and Resilience in Adolescence project of the Oxford Mindfulness Centre, the goal is to use schools and teachers to equip students with the "ability to "respond not react" in emotional stressful situations. That is to say, being able to step back from over-learned automatic ways of reacting and choose to respond more resiliently" (Oxford Mindfulness Centre, 2016). Resilience should be understood in this respect as an avoidance of extreme, particularly negative emotions (Robertson, 2012). Toward achieving this goal, mindfulness is promoted as a set of practices that teach how to cope with distressful situations and the emotions associated with them, providing practitioners with a basis for a more even and reflectively modulated emotional life (Kabat-Zinn, 2012). Given these objectives and the terms in which mindfulness is cast, it is not surprising, then, that we see an adoption in education of mindfulness practices and a turning to ancient Western and Eastern philosophies (Wren, 2014; Cen and Yu, 2014; Roeser *et al.*, 2014). In both West and East, we find a historical focus on self-control and a well-regulated and organized emotional life (for an example of exceptionally strong parallels in this respect, see Kuzminski, 2008), a focus that resonates well with the objectives of the policies, education programs and trends under consideration here.

Ancient Western philosophy, as it is increasingly being appreciated, was less an effort to provide coherent arguments about the universe than an effort to provide philosophies for living, tools for happy, virtuous and well-regulated emotional lives (Hadot, 2002; Nussbaum, 2009). Modern forms of evidence-based therapies, cognitive behavioral therapy or those therapies closely related to it, are deeply indebted to ancient Western philosophy both in the way they conceive of the relationship between reflection and emotion (Cooper, 1999) and in the value given to lives lived relatively free from irrational and extreme emotions (Robertson, 2010). The proper moral life is one in which people have ordered and structured themselves, informed by the latest and best science (Gill, 2009). Ethical and moral education turns to ancient Western philosophy today precisely to recuperate this therapeutic and reform-minded focus (Wren, 2014).

More obviously relevant with respect to mindfulness, however, is ancient Eastern philosophy and spirituality, particularly as it centres on contemplative practices (Kabat-Zinn, 2012). Concerned, as we are, with neuroeducation in North America and the UK, the question becomes: Why is an Eastern practice, such as mindfulness, being introduced today as an educational practice in the West? A full answer to this question would deal with a multiplicity of contributing factors. However, it seems likely that mindfulness is being practiced more generally across North America in response to a "broad antimodernist lament" (Harrington, 2008: 206) that characterizes a particular mood in

North America and Europe. As historian Anne Harrington notes, this lament stipulates that "[t]he stress of our modern way of life in the West … has damaged our hearts … and made us far more unhappy than we had ever imagined we would be. Nevertheless, there is hope for us" (Harrington, 2008: 206). Hope, in this case, comes from the East because "Easterners" have supposedly not adopted the same stressful and dehumanizing practices that "Westerners" have. Harrington encapsulates this dichotomy: "We are modern; they have stayed in touch with ancient traditions. We are harried and tense with stress; they speak and act wisely from a place of deliberate repose and contemplation" (Harrington, 2008: 206). This dichotomy, of course, is but a reiteration of a centuries old narrative in which some in the West view their own civilization as morally corrupt and culturally bankrupt. If one subscribes to this narrative, there are two solutions to this state of affairs: either turn to the past (as the Romantics famously did) or turn to the East (as so many Orientalists have and continue to do) in order to find resources to address Western shortcomings. In either case, whether it is through turning to the past or turning to the East, what is sought are traditional ways of dealing with a very modern problematic; namely, how one can live a relatively stable emotional life in times that are considered to be turbulent, stressful and uncertain (Harrington, 2008).

As mentioned above, some scholars see the adoption of therapy-informed interventions in education as evidence of a larger trend they identify as the "therapeutic turn" in North America and the UK (Ecclestone and Hayes, 2009; Furedi, 2003; Nolan, 1998). Indeed, the American psychologist Albert Ellis famously observed that he believed that the future of psychology and psychotherapy was to be found in the school system. In order to better understand the move to introduce therapy-based ideas and practices into schools, in this case mindfulness, we need to understand the way that mindfulness as a practice has been validated by neuroscience over the past several decades, the way that mindfulness is promoted as a practice for teaching emotional self-regulation, and the way that childhood and adolescence is framed as periods of vulnerability with respect to the vagaries of modern life. Before returning to the former two factors, we deal now with the latter.

Adolescents: broken by modern life; brains in flux

While neuroeducation is still very much an effort to make pedagogy more effective in terms of teaching competencies that are dictated by standards committees and future potential jobs, as we have attempted to illustrate above, the adoption of mindfulness and other neuroscience-validated practices must be understood in conjunction with a larger societal trend and the demands it creates. In particular, we stress the anxiety many policy-makers and educators express in North America and the UK concerning pupils and their ability to adapt to modern challenges and pressures. Critics of the so-called therapeutic turn in education actually see this concern as a result of therapeutic education itself. As they see it, therapeutic education disempowers, not empowers, students:

> The logical conclusion of being immersed in popular therapy is to see ourselves as all suffering, to a greater or lesser extent, from the negative emotional effects of diverse life experiences and events, where emotional survival of the everyday trials and tribulations of life become the ultimate goal. These ideas are reinforced by media and political attention to a plethora of fears about the impact of modern life on how people feel about themselves and their lives.
>
> (Ecclestone and Hayes, 2009: 5–6)

The adoption of mindfulness in the classroom, then, is a result of broad-based concern about, in our case, children and adolescents, and it is a therapeutic form of psychological governance which is developed to address these collective concerns.

Indeed, adolescents are frequently framed by scientists, policy-makers and the media as particularly vulnerable to the threats of modern life. Although since the early 1900s, with G. Stanley Hall's now legendary description of adolescence, Western society has viewed teenagers as both vulnerable and dangerous (Lesko, 2012), new moral panic about risks to young people's mental and physical health continue to capture the attention of society. Anxieties around substance abuse, risky driving, teenage pregnancy, digital addiction, distractedness or predatory gangs have become the substance of sensationalized news pieces, but also the focus of neuroscientific research in terms of the developing teenage brain, and are broken down into questions related to, for example, immature impulse control or underdeveloped empathy, each of which depend on the structural and functional development of the prefrontal cortex, the last area to develop in the teenage brain (Choudhury *et al.*, 2015).

Moreover, the line between healthy and disordered is increasingly blurred in adolescent brain research. Adolescence is understood as a paradoxical period of brain development during which teens undergo a phase of inevitable and normal "pathological" behavior as a result of their "disorganized," "underdeveloped" or "maturing" cortex simultaneously with increased risk for the onset of mental illness (Dahl, 2004). In other words, the typically developing brain is seen as "at risk." Thus, not only is adolescence problematized in terms of a period of vulnerability to modern threats, but it is also seen as inherently risky by virtue of an underdeveloped brain and, at the same time, as particularly receptive to carefully designed interventions. Understood to be a new "sensitive period" of brain development, this plastic, or malleable, period of developmental flux serves as a golden opportunity to guide normal development or remedy aberrant processes (Choudhury and McKinney, 2013). Neuroscience thus reframes older worries about how to guide the development of young people at this liminal and unstable moment of development while providing a substrate of hope and a mechanistic process of possibility and potential through the model of the plastic teenage brain (Choudhury and Moses, in press).

Plastic brain as ethical substance

We argue that the link we are making between developments in neuroscience and a concern for young students and effective pedagogical practices is strengthened through the idea of the brain as an ethical substance to which educators and children themselves are made to feel responsible for giving shape and form. If we are correct in drawing parallels between the moral and emotional objectives of ancient Western and Eastern philosophies and the moral and emotional objectives of neuroeducation in general – and in particular, the adoption of mindfulness – we must make one sharp distinction between those philosophies and contexts and the present one. Ethical substance, as Foucault (1990) theorized it, is the material part of the self one works on in order to craft oneself into a moral and well-regulated person. To give one example, Foucault identifies the flesh as the relevant ethical substance of medieval Christians. The flesh was the real and metaphorical material that needed to be worked on in order fashion oneself as a Christian. We are suggesting that for neuroeducation, the brain provides that substantive and metaphorical material that pupils learn to craft through practices like mindfulness.

In neuroeducation, neural pathways are the real and imagined site of relevant moral and pedagogical interventions. Like the flesh, the brain is both real and an image that is proposed and reproduced through brain-based interventions. What makes the brain such a fruitful site of intervention is that it is located at the threshold, intimately connected as it is to the senses, between the body and its social networks and environments, and that it is plastic, changeable, malleable. The brain is a site of encounter amenable to manipulation. This is what is usually meant by both the scientific and popular use of the term neuroplasticity. As we attempt to illustrate above, neuroscience, and the reports of neuroscience in the media, have given us the notion of children and adolescents as individuals with particularly plastic brains. Neuroeducation, then, becomes an effort at prevention and early intervention to stave off the worst of feared outcomes such as depression and severe emotional dysregulation. More to the point, though, the objective of teaching pupils to "respond not react" to stressful, emotionally charged situations has put the brain (Oxford Mindfulness Centre, 2016) and, in particular, the prefrontal cortex as the location of "executive function" – as that which must be crafted in order to achieve the most positive outcomes (Siegel, 2012).

As we illustrate through the example given in the next section of this chapter, by keeping in mind a particular image of the desired impacts to the brain while engaging in mindfulness or another brain-based practice, a virtuous feedback loop is formed between the ethical practice and what is imagined as being done to the brain. Rarely can a person know if any change has in fact taken place to the brain, so she must reassure herself by looking to the neuroscience and confirming the effectiveness of the practice by listing a series of benefits she is experiencing. The material reality of the brain, its plasticity and its location at the interface between individuals and their environments all contribute to what is seen as the effectiveness of the practice (e.g. mindfulness). What goes unnoticed, however, is the difference between the brain as a real organ in people's heads and the brain as an ethical object that motivates new practices and behaviors.

A case in point: a hand model of the brain

Educators, psychiatrists, parents and young people, particularly in the US, increasingly cite brain-based manuals as influential in their behavior management models. In recent ethnographic work that we conducted, popular science manuals of Daniel J. Siegel were particularly sought after. Through these manuals, along with workshops and online guides, Siegel has promoted the use of a hand model of the brain as a pedagogical tool in the service of the mindful regulation of emotions.

If you hold your forearm and hand vertically, place your thumb across your palm and then wrap your fingers over and around your thumb, making a fist with the thumb at the centre, you create a model of the brain, explains Siegel, where the cerebral cortex is represented by the fingers, the limbic area by the thumb and the brainstem by the palm (Siegel, 2012). The purpose of teaching the hand model, according to Siegel, is to help people, in our case children, learn the skill of consciously "integrating" brain processes. As we demonstrate below, the integration of brain processes is an objective of mindfulness more generally, including older and more directly Buddhist-inspired mindfulness practices. Siegel writes: "Integration in the brain creates a balanced and coordinated nervous system. … An integrated brain is important for a resilient and healthy mind" (2012: 3–3).

In the hand model, the prefrontal cortex is represented by the two middle fingernails. These are tucked between the thumb and the palm, representing the limbic areas and the brainstem respectively. Their central location in the model reflects their key role in what Siegel identifies as an integrated and well-regulated brain. Siegel explains:

> Notice how when your fingers enfold your thumb, this middle prefrontal region touches the limbic and brainstem areas, just as it is directly linked to these regions in the brain itself. If you lift your fingers up and put them back down, you may be able to see how this region is integrative – it links widely separated areas to each other.
>
> (2012: 10–5 to 10–6)

Integration is important so that the prefrontal areas can more effectively regulate; that is, exert a dominating influence by "modulating input from the lower areas" (Siegel, 2012: 3–4). The brainstem, the amygdala and the body itself can generate strong and overwhelming impulses. The cortex in this case, especially the prefrontal regions, can exert an inhibitory influence on these impulses – a process called "descending inhibition" – even overriding "sub-cortical activations" (Siegel, 2012: 3–4). This is the neural process by which someone regulates their emotions.

When the prefrontal cortex does not perform its inhibitory and integrative role, however, people can become dysregulated or, as Siegel puts it colloquially, "flip their lid." This situation is reflected in the hand model by straightening out the fingers and removing them from contact with the thumb and the palm of the hand: "Raise your fingers up suddenly and you'll have a visual image of how this *low road* behaviour can occur. Without the modification of the inte-grative functioning of the prefrontal cortex, the lower and more impulsive limbic brainstem areas can run amok" (Siegel, 2012: 10–6). That is, when the prefrontal cortex does not play its modulating role, "'emotional' or 'amygdala' hijacking" can occur – the very situation that the teaching of the hand model is aiming to ameliorate (Siegel, 2012: 10–6).

The purpose of the hand model, as with other mindfulness-based practices, is to help students to pause and strengthen integrative and regulative brain pro-cesses. Proponents, including Siegel and Rock, argue that this occurs by noticing, labeling and focusing attention on, in this case, the relevant brain regions and processes. As Rock puts it, "[w]hen you change your attention you are ... faci-litating 'self-directed neuroplasticity.' You are rewiring your own brain" (2009: 209). Instead of being overwhelmed by subcortical emotions and activations, children and adolescents can identify what is occurring in their brains, make helpful decisions (such as, for example, "taking time out") and strengthen integrative and regulatory processes. As Siegel puts it, "[s]chools that teach this model empower their pupils to know how their own internal mechanisms work so that they can be more aware of their impulses and become more flexible in their responses" (Siegel, 2012: 10–6). In this respect, the hand model differs from more traditional mindfulness practices, like "following" one's breath, by making an explicit link with the brain as the relevant ethical substance to be transformed and shaped. Although promoters and teachers talk about how mindfulness meditation will "strengthen" students' capacity to pay attention and regulate their emotions, the ethical molding in this case is less directly focused on the ethical substance itself. The hand model, on the other hand, makes a more direct link between labeling and focusing one's attention and the work that is being done on the brain as an ethical substance.

Siegel's and Rock's views on mindful practices, the focusing of attention and the transformation of brains, as we have glossed them, are not only of particular relevance because they are popular proponents of brain-based interventions; they are also pertinent because both men are on the board of directors at the Blue School in New York, one of the schools that has imple-mented mindfulness and other brain-based pedagogical practices in their grade school curricula and at which one of us has conducted some interviews.

A teacher who was interviewed by one of us for a previous study described the effectiveness of the hand model: "young kids now are talking about how their amygdala [are getting] overactive. I think that's good. In a way that level of critical distance and spaciousness between one part of mind and another can be very powerful" (Choudhury and Moses, in press). Here, we can see how this teacher makes an allusion to the idea that a mindful practice involves one part of the brain observing another. The highly portable hand model of the brain facilitates this observation by making a model of the brain readily available, putting it directly in front of the person, allowing them to see, whenever they need to, a representation of the relationship between the prefrontal cortex and the subcortical brain. In our language, the hand model makes the ethical substance and work on the self visible and explicit.

By talking about their "amygdala" getting "overactive," children focus their attention in such a way as to alter their brain and increase their regulation of subcortical impulses (Choudhury and Moses, in press). As a promoter and board member of the Blue School told one of us even more emphatically:

> when you've flipped your lid, kids can feel what that feels like. ... We've had kids as young as five years old come to their teacher with this model and say "I need to take a break, because I'm going to do things I don't want to do. I need a timeout."
>
> (Choudhury and Moses, in press)

The hand model helps people be able to more clearly observe themselves, label their emotions and cope with emotionally intense situations. By adopting mindful practices, children and adolescents can learn to live more mindfully, perhaps even changing who they become. As the teacher mentioned in the previous paragraph stated, "It's an interesting question about how [the hand model] interacts with developmental process and [children's] sense of self" (Choudhury and Moses, in press). Like mindfulness more generally in neuroeducation, the hand model is adopted as a pedagogical tool on the premise that it can help people to better mold their plastic brains in a way as to confront the uncertainties of modern existence with more resiliency.

Like with "following" the breath in one type of mindfulness meditation, the hand model "grounds" the self-shaping practice (Kabat-Zinn, 2012). By making the brain visible, putting it front and centre, the hand model keeps clearly in mind the substance to be molded, the site of behavioral intervention. This is particularly true if we talk about the middle two fingernails as the prefrontal cortex and its regulating function with proximate regions. When the prefrontal cortex is overloaded and cannot carry out its regulating role, the "lid is flipped": the fingers lift away from the thumb and stick straight up, illustrating the lack of "integration" in the brain. As with philosophies from the East and West, what is most important for the development of effective emotional regulation are practices, the tools and techniques and their repeated implementation. Whereas ancient philosophers and spiritual individuals might have naturalized their practices of self-control and emotional regulation by thinking of pre-modern substances such as "the flesh," after decades of investments in neuroscientific research into the brain and the popularization of neuroscientific knowledge, some people are repurposing old ethical practices and reattaching them to the brain as a substance to be shaped by attention and effort direct towards the better modulation of emotions.

Conclusion

Positioned at the intersection of developmental cognitive neuroscience and medical anthropology, we have described the concepts and modes facilitating the transfer of findings from cognitive neuroscience to educational practice and policy in the emerging area of mindfulness education. Despite the wide-spread uptake and influence on popular discourse, there are several important methodological limitations, epistemological dilemmas and ethical questions surrounding the translation of new brain data into areas of society involving the health and management of adolescent behaviors. Given the increasing relevance of brain models for various segments of society that determine health and social outcomes for adolescents, it is of particular interest to understand how practitioners and policy-makers in various arenas of social life (mental health, education and law) interpret, utilize and translate these new neurobiological models of adolescent behaviors and illness.

The mindfulness project in neuroeducation is exemplary of ways in which knowledge of the brain provides an evidence base for programs aimed at maximizing well-being in various domains of adolescent life in ways that can enable optimal "adaptive and successful" development (Paus, 2005) directed towards the idea of productive adult citizens (Kirkwood *et al.*, 2008). In particular, the period of adolescence, in both good and poor mental health, represents a decisive stage of development that is especially receptive to various forms of intervention and environmental stimuli, one that should be protected to increase cognitive capacities or "mental capital" for the future. The brain's plasticity over the lifespan, and the particular malleability of the prefrontal cortex during adolescence, functions as an explanation for the increased likelihood of risk-taking, antisocial behavior and mental illness. At the same time, this very notion of plasticity serves as the backbone for learning and developmental potential and the desired outcomes of mindfulness training among adolescents. Adolescents – with their malleable brains and at this liminal, vulnerable and impressionable stage of growth – are ideal candidates for mindfulness-based interventions.

The fervent enthusiasm around mindfulness, a practice derived from Buddhist philosophy, is in many ways surprising for contemporary educational policy in the West and especially among neuroscientists. We have attempted to address possible reasons for this coupling, pointing to a wider cultural dis-satisfaction in which "Eastern" ideas have found new resonance in the "West" and, in particular, within Western biomedicine (Harrington, 2008). In this context, mindfulness meditation as an evidence-based practice linked to state-of-the-art neuroscience provides hopeful yet secular horizons for education and for the future of young people in a doubly progressive way: not only does it import values of compassion and self-regulation and prioritize social-emotional development over academic achievement and competition, it also promises the cultivation of these virtues through visualizable, measurable and fundamental processes in the seat of self-control in the brain. By recruiting the latest research on neuroplasticity, attention and emotional regulation, policy-makers and educators are putting forth the developing brain as a site of psychological governance. While the objectives of achieving emotional regulation and greater resilience may be very old, even ancient, the brain, as a real and plastic organ and as an ethical substance, gives an entirely modern basis for hope and enthusiasm.

References

Ansari, D., Coch, D. and De Smedt, B. (2011) "Connecting education and cognitive neuroscience: where will the journey take us?" *Educational Philosophy and Theory*, 43(1): 37–42.

Battro, A. M., Fischer, K. W. and Léna, P. J. (Eds.) (2008) *The Educated Brain: Essays in Neuroeducation*. Cambridge University Press, Cambridge.

Blakemore, S. J. and Frith, U. (2005) *The Learning Brain: Lessons for Education*. Blackwell, Oxford.

Blakemore, S. J. and Choudhury, S. (2006) "Development of the adolescent brain: implications for executive function and social cognition." *Journal of Child Psychology and Psychiatry*, 47(3–4): 296–312.

Bruer, J. T. (1997) "Education and the brain: A bridge too far." *Educational Researcher*, 26(8): 4–16.

Bush, G. (1990) Presidential Proclamation 6158 – Decade of the Brain, 1990–1999. July 17.

Cantlon, J. F., Brannon, E. M., Carter, E. J. and Pelphrey, K. A. (2006) "Functional imaging of numerical processing in adults and 4-y-old children." *Plos Biology*, 4(5): e125.

Cen, G. and Yu, J. (2014) "Traditional Chinese philosophies and their perspectives on moral education," in L. Nucci, D. Narvaez and T. Krettenauer (Eds.) *Handbook of Moral and Character Education*, second edition. Routledge, London, pp. 30–42.

Changeaux, J. P. (1985) *Neuronal Man*. Pantheon, New York.

Choudhury, S. (2010) "Culturing the adolescent brain: what can neuroscience learn from anthropology?" *Social Cognitive and Affective Neuroscience*, 5(2–3): 159–167.

Choudhury, S. and Slaby, J. (2012) *Critical Neuroscience: A Handbook of the Social and Cultural Contexts of Neuroscience*. Wiley-Blackwell, Chichester.

Choudhury, S. and McKinney, K. A. (2013) "Digital media, the developing brain and the interpretive plasticity of neuroplasticity." *Transcultural Psychiatry*, 50(2): 192–215.

Choudhury, S. and Moses, J. M. M. (in press) "Mindful interventions: youth, poverty and the developing brain." *Theory and Psychology*.

Choudhury, S., McKinney, K. A. and Merten, M. (2012) "Rebelling against the brain: public engagement with the 'neurological adolescent'." *Social Science and Medicine*, 74(4): 565–573.

Choudhury, S., McKinney, K. and Kirmayer, L. J. (2015) "'Learning how to deal with feelings differently': Psychotropic medications as vehicles of socialization in adolescence." *Social Science and Medicine*, 143: 311–319.

Claxton, G. (1998) *Hare Brain, Tortoise Mind. Why Intelligence Increases When You Think Less.* Fourth Estate, London.

Cooper, J. (1999) *Reason and Emotion*. Princeton University Press, Princeton, CT.

Dahl, R. E. (2004) "Adolescent brain development: a period of vulnerabilities and opportunities. Keynote address." *Annals of the New York Academy of Sciences*, 1021, 1–22.

Ecclestone, K. and Hayes, D. (2009) *The Dangerous Rise of Therapeutic Education*. Routledge, London.

Farah, M. J. (2005) "Neuroethics: the practical and the philosophical." *Trends in Cognitive Sciences*, 9(1): 34–40.

Fischer, K. W. (2009) "Mind, brain, and education: building a scientific groundwork for learning and teaching." *Mind, Brain, and Education*, 3(1): 3–16.

Foucault, M. (1990) *The Use of Pleasure: Volume 2 of the History of Sexuality* (trans. R. Hurley). Vintage Books, New York.

Furedi, F. (2003) *Therapy Culture: Cultivating Vulnerability in an Uncertain Age*. Routledge, London.

Giedd, J. N. and Denker, A. H. (2015) "The adolescent brain: insights from neuroimaging," in J.-P. Bourguignon, J.-C. Carel, and Y. Christen (Eds.) *Brain Crosstalk in Puberty and Adolescence*. Springer International, Cham, pp. 85–96.

Gill, C. (2009) *The Structured Self in Hellenistic and Roman Thought*. Oxford University Press, Oxford.

Gonon, F., Bezard, E. and Boraud, T. (2011) "Misrepresentation of neuroscience data might give rise to misleading conclusions in the media: the case of Attention Deficit Hyperactivity Disorder." *Plos One*, 6(1): e14618.

Goswami, U. (2006) "Neuroscience and education: from research to practice?" *Nature Reviews Neuroscience*, 7(5): 406–413.

Hadot, P. (2002) *What is Ancient Philosophy?* (trans. M. Chase). Harvard University Press, Cambridge, MA.

Haidt, J. (2006) *The Happiness Hypothesis: Finding Modern Truth in Ancient Wisdom*. Basic Books, New York.

Harrington, A. (2008) *The Cure Within: A History of Mind-body Medicine*. W. W. Norton and Company, New York.

Hillman, C. H., Erickson, K. I. and Kramer, A. F. (2008) "Be smart, exercise your heart: exercise effects on brain and cognition." *Nature Reviews Neuroscience*, 9(1): 58–65.

Howard-Jones, P. A. (2009) "Scepticism is not enough." *Cortex*, 45(4): 550–551.

Hyland, T. (2011) *Mindfulness and Learning: Celebrating the Affective Dimension of Education*. Springer, London.

Illes, J., Moser, M. A., McCormick, J. B., Racine, E., Blakeslee, S., Caplan, A. and Weiss, S. (2010) "Neurotalk: improving the communication of neuroscience research." *Nature Reviews Neuroscience*, 11(1): 61–69.

Kabat-Zinn, J. (2012) *Mindfulness for Beginners: Reclaiming the Present Moment – and Your Life*. Sounds True, Boulder, CO.

Kahneman, D. (1973) *Attention and Effort*. Prentice-Hall, New Jersey.

Kirkwood, T., Bond, J., May, C., McKeith, I., Teh, Min-Min (2008). Foresight Mental Capital and Wellbeing Project: Mental Capital Through Life, Future Challenges. The Government Office for Science, London.

Knudsen, E. (2004) "Sensitive periods in the development of the brain and behaviour." *Journal of Cognitive Neuroscience*, 16(8): 1412–1425.

Kuzminski, A. (2008) *Pyrrhonism: How the Ancient Greeks Reinvented Buddhism*. Lexington Books, Lanham.

Learning2Breathe (2016) "Learning2Breathe: A Mindfulness Curriculum for Adolescents" [online]. *Learning2Breathe*. Available at: http://learning2breathe.org/about/why-strengthen-emotion-regulation-skills-in-adolescents

Lesko, N. (2012) *Act Your Age: A Cultural Construction of Adolescence*. Routledge, New York.

McCabe, D. P. and Castel, A. D. (2008) "Seeing is believing: the effect of brain images on judgments of scientific reasoning." *Cognition*, 107(1): 343–352.

Markram, H. (2012) "The human brain project." *Scientific American*, 306(6): 50–55.

Maxwell, B. and Racine, E. (2012) "The ethics of neuroeducation: research, practice and policy." *Neuroethics*, 5(2): 101–103.

Meiklejohn, J., Phillips, C., Freedman, M. L., Griffin, M. L., Biegel, G., Roach, A., Frank, J., Burke, C., Pinger, L., Soloway, G., Isberg, R., Sibinga, E., Grossman, L. and Saltzman, A. (2012) "Integrating mindfulness training into K-12 education: fostering the resilience of teachers and students." *Mindfulness*, 3(4): 291–307.

Nolan, J. L.Jr (1998) *The Therapeutic State: Justifying Government at Century's End*. NYU Press, New York.

Nussbaum, M. C. (2009) *The Therapy of Desire: Theory and Practice in Hellenistic Ethics*. Princeton University Press, Princeton, CT.

OECD (2002) Understanding the Brain: Towards a New Learning Science. OECD, Paris.

Ortega, F. and Vidal, F. (2011) *Neurocultures: Glimpses into an Expanding Universe*. Peter Lang, New York.

Oxford Mindfulness Centre (2016) Mindfulness and Resilience in Adolescence: The MYRIAD project [online]. *Oxford Mindfulness Centre*. Available at: www.oxfordmindfulness.org/mindfulness-resilience-adolescence/

Park, A. and Sabourian, H. (2011) "Herding and contrarian behavior in financial markets." *Econometrica*, 79(4): 973–1026.

Paus, T. (2005) "Mapping brain maturation and cognitive development during adolescence." *Trends in Cognitive Science*, 9(2): 60–68.

Pickersgill, M., Martin, P. and Cunningham-Burley, S. (2015) "The changing brain: neuroscience and the enduring import of everyday experience." *Public Understanding of Science*, 24(7): 878–892.

Pitts-Taylor, V. (2010) "The plastic brain: neoliberalism and the neuronal self." *Health*, 14(6): 635–652.

Racine, E., Waldman, S., Rosenberg, J. and Illes, J. (2010) "Contemporary neuroscience in the media." *Social Science and Medicine*, 71(4): 725–733.

Raz, A. (2012) "Critical neuroscience: from neuroimaging to tea leaves in the bottom of a cup," in S. Choudhury and J. Slaby (Eds.) *Critical Neuroscience: A handbook of the Social and Cultural Contexts of Neuroscience*. Wiley-Blackwell, Oxford, pp. 265–272.

Robertson, D. (2010) *The Philosophy of Cognitive Behavioural Therapy (CBT): Stoic Philosophy as Rational and Cognitive Psychotherapy*. Karnac Books, London.

Robertson, D. (2012) *Build Your Resilience: How to Survive and Thrive in any Situation*. Teach Yourself, London.

Rock, D. (2009) *Your Brain at Work*. Harper Collins, New York.

Roeser, R. W., Vago, D. R., Pinela, C., Morris, L. S., Taylor, C. and Harrison, J. (2014) "Contemplative education: cultivating ethical development through mindfulness taining," in L. Nucci, D. Narvaez and T. Krettenauer (Eds.) *Handbook of Moral and Character Education*, second edition. Routledge, London, pp. 223–247.

Rose, S. (2005) *The Future of the Brain: The Promise and Perils of Tomorrow's Neuroscience*. Oxford University Press, Oxford.

Rose, N. S. and Abi-Rached, J. M. (2013) *Neuro: The New Brain Sciences and the Management of the Mind*. Princeton University Press, Princeton, CT.

Sarkari, S., Simos, P. G., Fletcher, J. M., Castillo, E. M., Breier, J. I. and Papanico-laou, A. C. (2002) "The emergence and treatment of developmental reading disability: contributions of functional brain imaging." *Seminars in Pediatric Neurology*, 9(3): 227–236.

Shaywitz, S. E. and Shaywitz, B. A. (2004) "Reading disability and the brain." *Educational Leadership*, 61(6): 6–11.

Siegel, D. J. (2012) *Pocket Guide to Interpersonal Neurobiology: An Integrative Handbook of the Mind*. W. W. Norton & Company, New York.

Tallis, R. (2004) *Why the Mind is Not a Computer: A Pocket Lexicon of Neuromythology*, second edition. Imprint Academic, Exeter.

Tallis, R. (2011) *Aping Mankind: Neuromania, Darwinitis and the Misrepresentation of Humanity*. Acumen Publishing, Durham.

Temple, E., Deutsch, G. K., Poldrack, R. A., Miller, S. L., Tallal, P., Merzenich, M. M. and Gabrieli, J. D. (2003) "Neural deficits in children with dyslexia ameliorated by behavioural remediation: evidence from functional MRI." *Proceedings of the National Academy of Sciences*, 100(5): 2860–2865.

Thomson, J. M., Leong, V. and Goswami, U. (2013) "Auditory processing interventions and developmental dyslexia: a comparison of phonemic and rhythmic approaches." *Reading and Writing*, 26(2): 139–161.

Weisberg, D. S., Keil, F. C., Goodstein, J., Rawson, E. and Gray, J. R. (2008) "The seductive allure of neuroscience explanations." *Journal of Cognitive Neuroscience*, 20(3): 470–477.

WellcomeTrust (2015) "Large-scale trial will assess effectiveness of teaching mindfulness in UK schools" [online]. *Wellcome Trust*, 16 July. Available at: www.wellcome.ac.uk/News/Media-office/Press-releases/2015/WTP059495.htm

Wilson, A. J., Dehaene, S., Dubois, O. and Fayol, M. (2009) "Effects of an Adaptive Game Intervention on Accessing Number Sense in Low Socioeconomic Status Kindergarten Children." *Mind, Brain, and Education*, 3(4): 224–234.

Wolpe, P. R. (2002) "The neuroscience revolution." *The Hastings Center Report*, 32(4): 8.

Wren, T. (2014) "Philosophical moorings," in L. Nucci, D. Narvaez and T. Krettenauer (Eds.) *Handbook of Moral and Character Education*, second edition. Routledge, London, pp. 11–29.

8 Behavioural science, randomised evaluations and the transformation of public policy

The case of the UK government

Peter John

Behaviour change policy conveys a powerful image: that of psychologists and scientists, maybe wearing white coats, messing with the minds of citizens, doing experiments on them without their consent, and seeking to manipulate their behaviour. Huddled in offices in the bowels of Whitehall, or maybe working out of windowless rooms in the White House, behavioural scientists are redesigning the messages and regulations that governments make, operating far from the public's view. The unsuspecting citizen becomes something akin to the subjects of science fiction novels, such as Huxley's *Brave New World* or Zamyatin's *We*. The emotional response to these developments is to cry out for a more humanistic form of public policy and a more participatory form of governance. The implication is that public policy should be placed firmly in the hands of citizens and their elected representatives.

Of course, such an account is a massive stereotype. Yet something of this viewpoint has emerged as a backdrop to recent academic work on the use of the behavioural sciences in government. Those working from a critical social science perspective observe the rise of 'the psychological state' because of a step change in use of psychological and other forms of behavioural research to design public policies (e.g. Jones *et al.*, 2013a, 2013b; Leggett, 2014; Strassheim and Korinek, 2016). This claim is inspired by sociological studies of science and government, which has been subject of much theoretical work in recent years. Drawing on the work of Foucault, the research programme has explored the practice of governmentality in the field of behaviour change (see Jones et al., 2013a: 182–8): the 'central concern has been to critically evaluate the broader ethical concerns of behavioural governance, which includes tracing geohistorical contingencies of knowledge mobilized in the legitimation of the behaviour change agenda itself' (2013a: 190). This line of work presents a subtle set of arguments and claims that an empirical account, such as one presented in this chapter, cannot – and should not – challenge. Nonetheless, it is instructive to find out more about the phenomenon under study and to understand how the use of behavioural ideas and randomised evaluations is limited and structured by the institutions and actors in the political process, which are following political and organisational ends. Of particular interest is the incremental and patchy nature of the diffusion of ideas and how the use of behavioural sciences meshes with existing standard operating procedures and routines of bureaucracies. The path of ideas in public policy is usually slow – one of gradual diffusion and small changes in operating assumptions – and this route is likely for the use of behavioural sciences. Ideas can still make progress, helped by the fragmented and decentralised policy process; but their evolution is neither predictable or uniform. Nudges do have the power to create innovations in public policies, but they need to be skillfully advocated by bureaucrats, politicians and talented outsiders. The implication of this line of argument is that agency as well as structure plays an important role in the adoption and diffusion of ideas from the behavioural sciences. It implies a more limited use of ideas and evidence than is implied by the critical writers in this field, and public argument and debate play a central role.

The structure of this chapter is as follows: the first section reviews the literature on the use of evidence in public policy in the UK and assesses the progress of ideas from the behavioural sciences, in particular their advocacy by the Behavioural Insights Team; then, the chapter discusses the advantages and challenges of trials to test for evidence; and finally, there is an assessment of the likely progress of behavioural public policy.

Evidence and policy in the UK

As an advanced economy with a strong research and higher education sector, the UK government has always paid a lot of attention to science and scientific forms of knowledge. This official interest goes back centuries and is reflected, for example, in the establishment and persistence of the Royal Societies. The Haldane Report of 1911 sought to enshrine the principle of scientific research in the organisation of government. The UK has used scientific evidence extensively, and this is often promoted by government (see Edgerton, 2009), which has directly employed a large number of scientists. Social sciences have been important as part of this framework of providing intellectual support. They were argued for strongly in the post-1945 era and have been successful and are seen as legitimate, especially after the formation of the Social Science Research Council in 1965. The way in which UK government works shows the importance of scientific and social science advice: civil servants responsible for public policy work alongside personnel following research careers in the Government Social Research Service or in Government Science and Engineering.

The use of evidence in public policy operates in a political and institutional arena. It needs to satisfy the political demands for policies and is deployed in a fragmented institutional context where government departments and other agencies are in competition with each other for resources and political attention. King's (1998) research on the origins of the Social Science Research Council (later renamed the Economic and Social Research Council) shows how the arguments for setting it up were strong because of the context of the time and the demands for neutral, value-free social science evaluation.

Research on the use of evidence to evaluate public policies finds competing demands for policies. Messages from evaluation have to compete with other sources of information (Weiss, 1972). Such findings do not show that research is ineffective, but that a simple model whereby the production of more convincing evidence leads to policy change is not supported because of the political way in which policy is formulated and the need to show impact in a short period of time. Some of the difficulties of securing the impact of evidence, even from a government that was very sympathetic to evidence-based policies, was seen during Labour's period in office from 1997, such as in reforming policies to give support to families with very young children via the Sure Start programme. Eisenstadt (2011) recounts the twists and turns of the policy and the sheer difficulty of finding good evidence at the right time. The study draws attention to the complexity of the policy process and how advice given by experts is not always heard.

However, it would also be wrong to conclude that the policy-making process is an evidence-free zone. In fact, ministers and civil servants are highly engaged with evidence, partly because fact-based arguments are essential in the policy process (Majone, 1989). They are needed to advance political careers and to ensure that a government stays in office. Policy-makers, operating in response to debates in the media and in front of legislative committees, need to win the argument as well as exercise power. Collecting evidence underlying the use of policies is important in creating a favourable political backdrop so that new measures can be justified and defended. The opposition can make hay with findings from a research institute that puts the government on the back foot.

Studies of the policy process show that ideas are important but that they emerge unpredictably. Famously, Kingdon's (1984) account of the policy process stressed the importance of entrepreneurs seizing the initiative at a crucial point in time. It is much the same with the use of academic ideas and evidence. They need powerful advocates as described in Haskins and Margolis' (2015) account of the growth of evidence-based social policies under the Obama presidency, which were perhaps lacking in the Sure Start example. These accounts of the use of evidence and deployment of ideas in public policy resonate with the development of behavioural sciences in governments. There is no march of the psychologists and other behavioural scientists to control the policies of government, but there is a confluence of multiple influences in a complex system of government and agencies wherein it is hard to assume a uniform influence of behavioural ideas and evidence. The following section traces the influence of behavioural sciences in government and the growth of forms of evaluation that test out these kinds of initiatives.

The behavioural sciences and government

The study of individual behaviour is central to social science as students of society and politics have often wished to explain behaviours that matter for other outcomes, such as mobility in order to seek employment, voting that sustains democracies, or savings that help people in retirement – behaviours are central to human activity and are affected by other things that can be studied, such as attitudes and beliefs and the role of public and private institutions. For public policy, governments have always sought to influence human behaviour, such as volunteering to serve in the armed forces; public information campaigns have had a long pedigree; and many aspects of public policy are designed to change behaviour, such as tax changes or road safety measures. Such attention to the design of policies while also thinking about their consequences for behaviour and policy outcomes has gradually increased over time as the knowledge base has developed.

At the same time as there being these forms of knowledge in particular sectors, one discipline had a more integrated account of behaviour, which has had a huge influence on public policy. Economics has always had a privileged place in circles of decision-making, coming from the importance of the economy to Western societies and also from the self-confidence of economists and their relative agreement on models and methods. Most of all, economics offers an account of decision-making and outcomes at different levels of analysis, whether it be the individual, firm, group or government, which are linked by micro-level assumptions about rational behaviour. The other strong feature of economics is that it produces recommendations of great clarity based on the rational decision-making model. It can tell governments how to achieve their ends by modelling a policy choice based on rational reactions to the costs and benefits of public policy. This argument was famously made by Becker (1968) in working out how much to punish someone for a crime: government can adjust the law to affect how criminals or potential criminals make cost-benefit calculations about whether to commit a crime or not.

One important feature of the behavioural revolution is that it happened in the heart of the intellectual empire. Kahneman and Tversky applied concepts and findings from psychology to understand decision-making and wrote about the biases that people have when faced with new items of information (Kahneman, 1973; Kahneman and Tversky, 1979; Kahneman *et al.*, 1982, which can be read more informally in Kahneman, 2011). Such ideas are mainstream to cognitive psychology (see Newell *et al.*, 2007), but applied to economics, they were revolutionary in that they questioned the unit assumptions of economic theory. In prospect theory, for example, people weigh losses more highly than gains, but this means that they will not be able to respond rationally to the signals the public authority provides, which undermines or modifies the way in which government policy is based on adjusting costs and benefits.

Behavioural economics has become accepted and part of the mainstream in economics in spite of its lack of fit with the core theory of the discipline. In part, this is because behavioural economists have not made an assault on economic theory but, instead, have been focused on a set of empirical problems. Behavioural economists have used advanced mathematics and standard econometric models to solve their research problems and have been published in the top journals in the field, which can be observed if one reads the *curriculum vitae* of behavioural economists such as Richard Thaler with his work on savings and pensions (e.g. see Benartzi and Thaler, 2004). Another example is work on financial markets that looks at herding behaviour using mathematical models (Park and Sabourian, 2011), which can be useful in analysing bubbles and crises in the financial system. A behavioural economics academic paper does not look that different from one written using classical models.

It took a while for the findings of behavioural economics to diffuse from the academy to policy. In the UK, think tanks and government departments became aware of this line of work and began to publicise it. For example, even the New Economics Foundation, which is in no way a right-wing think tank, published its seven principles of behavioural economics (Shah and Dawney, 2005), which was a synoptic literature review. The Prime Minister's Strategy Unit, working to Tony Blair, produced its *Personal Responsibility and Behaviour Change* (Halpern and Bates, 2004), which incorporated many findings from behavioural economics. As the momentum gathered, government research started to address issues with ideas from behavioural economics (see John and Richardson, 2012), coordinated at first in the Government Economic and Social Research team located in the Treasury. Other units started up; for instance, in Department of Environment, Food and Rural Affairs (Defra). It should be noted that psychologists were employed by these units but the focus was interdisciplinary and, of course, economists played a strong role. As the 2000s rolled on, interest from the centre continued, partly due to the personal curiosity of the Cabinet Secretary at the time, Gus O'Donnell, who was an academic economist and who encouraged government to commission the influential MINSPACE report that listed out the findings of behavioural economics (Dolan *et al.*, 2010), researched and written before the Conservatives entered office in coalition with the Liberal Democrats in 2010. What is striking about this publication is neither that it has an attractive acronym nor that it was the most downloaded report from the website of the prestigious Institute for Government with its remit to improve the practice and functioning of government, but that the text and, in particular, the footnotes drew directly on work published in economics and other journals, effectively routing that knowledge into mainstream policy-making. At the same time that policy-makers were starting to look at academic knowledge more closely, academics were starting to realise that the ideas from the behavioural sciences had a wider, more generalist audience. A series of academic books appeared that reviewed evidence and pulled together findings from across many fields and disciplines to illustrate the size of the knowledge base. One example is *Behavioural Public Policy* (Oliver, 2013a), which consists of essays and commentary from academics. Another is the longer and more comprehensive, if less digestible, *The Behavioral Foundations of Public Policy* (Shafir, 2013), a collection of essays exploring theory and practice, including contributions from academic psychology.

Another route for dissemination is the popular non-fiction book, which gets reviewed in the quality newspapers and magazines, appears in prominent places on airport news stands, and is aggressively marketed by publishers hoping to make money – as well as being disseminated by the authors themselves through their websites and social media feeds. There is a long tradition of academics writing these kinds of books; for example, *Freakonomics* (Dubner and Levitt, 2005) summarised insights from clever and sophisticated empirical econometric studies that showed that policy-makers were following the wrong policies. Thaler and Sunstein's *Nudge* (2008) was an exemplary example of the 'trade book' written in an accessible style where the authors (or probably one of the author's) made fun of themselves and used everyday examples to advance their case while still resting their claims on academic studies. There are relatively few references to academic psychology; in fact, the main message of the book is that nudges amount to economic insights provided with a strong dose of common sense.

The nudge unit and beyond

The setting up of the Behavioural Insights Team, a unit in the Cabinet Office working directly to the Prime Minister, in the summer of 2010 is a story that has been well told (John, 2014), not least by the director, David Halpern, in his book *Inside the Nudge Unit* (2015). It is a tale of small beginnings and the way in which the unit built on the gradual diffusion of ideas about behavioural sciences described above. It benefited from the backing of the Conservative leader and his advisors who were interested in behaviour change and social policy and in what could be done to create a more prosocial citizenry, the so-called Big Society. Gus O'Donnell was still in place as head of the civil service and was ready to propose the idea to the incoming government; and of course it had as its director the same David Halpern who had played such a critical role in generating enthusiasm for behavioural ideas when he was in the Strategy Unit under Labour.

It started with just seven members, taking up just about half of a shared long desk in the cavernous offices at 1 Whitehall. Most were career civil servants seconded from other units with Halpern as the only psychologist, joined later by Laura Haynes. Paul Dolan, a behavioural economist, was only a team member for a short while. The team could not be called a psychological unit as its approach was more practical – building relationships around Whitehall and its environs and getting a sense of excitement going. Although they were plugged into the academic community, with the Academic Advisory panel set up in 2011,[1] the approach was to take some appealing ideas and apply them to practical problems of government departments and other agencies. A long cast list of celebrity academics, such as the psychologist Dan Ariely, came through Whitehall to give talks and attend meetings, and they also provided proposals to test. Team members gave presentations to government departments around Whitehall and had many meetings with officials to discuss how behavioural ideas could be applied to policies being developed. From these interactions, some officials came forward with proposals for behavioural interventions that were developed in partnership with members of the team. The team was in effect providing a service to these departments that was free at first. Team members also did desk research on different options studied in the literature in order to identify ones to test. The approach was to convey behavioural ideas as clearly as possible. This was symbolised by the simplification of the MINDSPACE acronym (Dolan *et al.*, 2010), with its basis in the economics and psychology literatures, to the shorter and more straightforward EAST: Easy, Attractive, Social and Timely (Service *et al.*, 2014). In presentations and in their publications, the team stressed the straightforward nature of their interventions and how big problems often have simple solutions. For example, over measures to encourage household to install energy-saving devices in their homes, team members emphasised that the main reason for lack of adoption of insulation was that people did not want to clear their lofts, and uptake increased fourfold if households were offered free loft clearance (Behavioural Insights Team, 2012a). When presented with a slide showing a cluttered loft, audience members could not help smiling as they saw their weak selves in the example. The result was that public officials in Whitehall and elsewhere felt they owned such unthreatening findings and research, and they wished to apply these fashionable ideas in their own bailiwicks.

The team was very public about its work, and there is no sense it was a secret cabal unleashing a programme of control of citizens around Whitehall. It was interested in trailblazing the proof of the concepts. This was very much true of the work on speeding up debt recovery using behavioural insights, summarised in the Behavioural Insights Team (2012b) report on 'fraud, error and debt'. The report is a guide to the work that has been done, containing studies by the team as well as summaries from desk reviews. It very much spoke to the agenda of government departments, seeking to add value on the top of policies that had already been agreed. The changes it encouraged were not radical step changes in citizen compliance with government ideas and policies, but modest changes based on working with the grain of citizen preferences. Many behavioural interventions, although having their origins in academic work, were easily comprehensible; for example, personalising SMS text messages to encourage citizens to pay their court fines (Haynes *et al.*, 2013). This use of nudge is consistent with a more citizen-friendly form of governance whereby innovations are adopted that take the viewpoint of the hard-pressed citizens seeking to manage their lives. Nudge needs bureaucrats who understand how people think as opposed to the rolling out of interventions designed in a psychological lab (John *et al.*, 2011).

Like any small unit operating in the relatively decentralised context of British government, where the centre does not have absolute power and itself operates more like a feudal court than a commander and enforcer (see Burch and Holiday, 1995), the team needed to persuade other parts of government and then share in the glory. It worked in a competitive environment of credit-claiming and searching for ministerial attention, but it survived relatively unscathed in spite of the inevitable spats and turf wars. It may have benefited from a shift in focus by the centre of government toward delivery and implementation, which had happened under prime ministers Blair and Brown. The 1997–2010 Labour government's focus on delivery was epitomised by the employment of the arch-moderniser Michael Barber (2008). The implementation focus carried on under the coalition government (2010–2015), especially as it was hard to agree major policy objectives between the two political parties.

The Behavioural Insights Team achieved a positive impact in the media, getting a good press both from left-of-centre outlets (e.g. Benjamin, 2013) as well as from the right-wing press (e.g. Bell, 2013). What many of these write-ups on the unit show is that journalists started out with a story about manipulation and the role of psychologists using their expertise to trick citizens, but in the end, they were won over by the common-sense approach and added value of such techniques. There was a similar kind of reaction in the public commentary and assessment of nudge policies. Initially, there was some resistance to the use of nudge policies and the role of the Behavioural Insights Team, the argument being that such an approach was designed to prevent the use of strong tools of government. Critics believed that nudge was a covert way of justifying a retreat of the state and that it entrenched the role of the private sector in government in fields such as food labelling (House of Lords, 2011). Academics have made the same kind of argument: nudge was not strong enough to achieve effective behaviour change and a more robust budge would be a better approach (Marteau *et al.*, 2011). Yet behaviour change advocates have criticised the idea that nudge implies that behavioural interventions apply just to information campaigns and the manipulation of defaults, whereas in fact behavioural science can be applied more generally across the tools of government to inform how they work (see Oliver, 2013b). In other words, greater use of the behavioural sciences allows policy-makers to think about the informational environment of all their actions so that 'all tools are informational now' (John, 2013a).

In short, the team has been a success and has been adroit in managing its external reputation. Its good fortune depended on responding to the political agenda and working with government departments and agencies. It mainly tweaked existing policies and implementation procedures rather than imposing an integrated vision. Above all, it carried out its task in a way that is public and transparent. Its record seems a long way from the image of the psychological state the chapter started with and more about the way in which a relatively small team has been able to craft useful improvements to standard procedures and helped public officials get better traction on how policies are implemented. The team publicised their policies to encourage more debate and take-up of behavioural insights. Such a demand-led approach has continued since the unit was spun out of government in 2014. It has attracted a range of clients with the same approach of information redesign. It has expanded its staff base to about 80 members. Although the team employs social psychologists, the modal background is economics; Michael Sanders, its director of research, is a behavioural economist. Meanwhile, the behavioural agenda continues in government with the formation of a behavioural economics unit within HM Revenue and Customs (HMRC) – the Behaviour Change Knowledge Network – with good links across government. Behavourial ideas diffuse to other jurisdictions, including to local government with local councils redesigning the considerable amounts of information and regulations they convey to citizens, such as local tax reminders (see Blume and John, 2014), or encouraging channel shift to the online delivery of services (John and Blume, 2015). The behavioural agenda works well in a decentralised arena where agencies focus on delivery and relationships with customers and citizens. Agencies can try out different ideas and customise their messages and regulations in different contexts.

The success of the Behavioural Insights Team and other units takes a path of gradual diffusion, whereby enthusiasm can play its role in getting buy-in from other agencies and groups. Outside the UK, a similar pattern emerged: units have largely been set up in the team's image, whether it be Behavioural Insights in the New South Wales government in Australia or the White House Social and Behavioral Sciences Team set up by President Obama. There is a similar unit operating in the German Federal Government. In some places, nudge units do not find favour, such as with the Australian federal government: proposals were put to the incoming Liberal administration in 2013 (John, 2013b) but the government did not, at first, set up a nudge unit or equivalent. In France, the employment of Oliver Oullier in the Centre for Strategic Analysis of the Prime Minister seemed to indicate the value of a cognitive behavioural scientist at the centre of government. But nudge ideas were ridiculed in the French press, and the subsequent administration under President Hollande did not go down the behavioural policy route. Behavioural policy has diffused in ways that depend on context and willingness to deploy these ideas. Nudge needs to be seen as feasible and acceptable by the elected politicians and elites in a society or else it is not going to work or even be adopted. It needs champions to push it forward, and as a project, it has to have good publicity and relevance for behavioural initiatives. In this way, the progress of behavioural ideas is political in the wider sense of that term, in that ideas need to work alongside political interests, and nudge advocates need to respond to the agenda of existing holders of power. Nudge then is not going to transform the state with a new set of ideas that can be used by authoritarian politicians seeking greater control and compliance over citizens; rather, it works within the fragmented and complex structure of the state where advocates are seeking to advance new ideas and to get the attention of senior policy-makers and those responsive for the delivery of policies. Nudge has been successful because it has worked within the existing agenda of state policies and according to the standard operating procedures of the bureaucracy. It appeals to the existing aims of bureaucrats and to those wanting more efficiency and traction from public policies.

Randomised evaluations

The use of behavioural sciences in government has been tested and promoted through the use of randomised controlled trials. This is a research method that randomises allocation of a person (or a unit, such as a local area) to a treatment group and then compares the outcomes of the treatment with those of a control group of similar people who did not get the treatment (or who received another treatment). A trial is not the only way to test for the efficacy of behavioural interventions; observational data can be used to find out why an intervention might work, especially when it has been gathered using quasi-experimental methods (e.g. matching subjects). Qualitative methods can also be important, especially in finding out why people might respond to a behavioural cue or not. Moreover, randomised evaluations have a number of drawbacks. They are quite complicated to carry out in public policy contexts because they need a high degree of control over the delivery process to ensure that randomisation happens cleanly, data is matched without errors, and the trial does not suffer from excessive attrition of its subjects (see a review in John, forthcoming). They often need to be done at specific times, in particular locations, and with people who are available. In such cases, there are difficulties in making inferences to other places and time periods. Trials work well when they are carried out in a range of locations and where it is possible to repeat the intervention many times, which of course is the conventional wisdom for trials of medical procedures and treatments. But usually with public policy, there is only one opportunity to test the intervention before it is rolled out, and it is hard to persuade policy-makers to test over and over again. With these complicated features, it is understandable why randomised controlled trials did not initially become a widely used method for evaluation even though their superiority in terms of making inferences or generalising has been known about since the 1920s. The statistical theory behind randomised controlled trials was developed by Fisher, based on the practical experience of advising on the agricultural experiments at Rothamsted. Although Fisher's *The Design of Experiments* was first published in 1935, the theory and practice had developed more gradually over the early years of the twentieth century. It was interesting how quickly contemporary policy-makers saw the potential of trials; for example, in testing for the effects of free school milk in Lanarkshire ('Student' 1931), even though this experiment was marred by implementation errors and challenges.

In spite of the early interest in field experiments, they took a long time to become established outside medicine and testing of health procedures. They did not really get going until the 1960s with experiments on tax and welfare in the US, which gradually developed into a corpus of work testing federal and state employment and welfare initiatives (Gueron and Rolston, 2013). In the US, field trials in public policy started to grow exponentially in the twenty-first century, in particular with large increases in trials linked to education policy and practice (see Haskins and Margolis, 2015). The other main area where trials are used to evaluate public policy is development with, for example, the Abdul Latif Jameel Poverty Action Lab and Innovations for Poverty Action. Development economists see the trial as a key way to evaluate interventions (see Duflo *et al.*, 2006).

The use of trials in the UK has been patchy. In terms of understanding and acceptance of trials, it has always been known that they have a number advantages that make them a preferred or desired form of evaluation, as seen, for example, in *The Magenta Book* (H M Treasury, 2011). But government departments had never been champions of trials, with the exception of employment where randomised interventions were introduced as early as 1991 (White and Lakey, 1992). But many experiments have not gone so well, often marred by political controversy and complex design issues that have weakened their impact and legitimacy. One example was the large trial carried out between 1998 and 2006 to evaluate whether the culling of badgers would reduce the spread of bovine Tuberculosis; this found no evidence that shooting the badgers worked and even indicated that it had the opposite effect of spreading the disease to other locations (Independent Scientific Group on Cattle TB, 2007). But the research generated a massive controversy and there was enough confusion about the design of the trial to ensure it did not influence policy (Dunlop, 2013).

Behavioural interventions are an exception to the general reluctance to test policies with trials. The key advantage is that prior research produces a number of recommendations that can be applied to relatively discreet stages or parts of the delivery process. Consider, for example, a favourite of behavioural interventions: the redesign of a letter to encourage someone to settle up a payment, which has been used by HMRC to encourage early payment of debts (Hallsworth *et al.*, 2014). The discretion of the policy-maker is in the wording of these letters and whether different phrases can be included alongside the normal and standard reminder text. The wording can be changed without recourse to legal advice. It is possible to give the instructions to send different letters (thus creating different treatment groups) and then examine the payment accounts of those who have been written to against identifiers that say which treatment group they are in. Tables can be produced showing payment rates according to treatment allocation, and standard tests of significance can be run based on the average payment or proportions in each group. The key is to word the different letters so that there can be comparison of feasible alternatives. As discussed earlier, these trials do not provide a pure test of a psychological mechanism that would stand peer review in a psychological science journal, which could only be delivered in a laboratory setting; they are simply tests that assess actions that are practical for governments in particular situations. Often the trial aims are decided in an interaction between the client and the researcher, each bouncing ideas back and forth. The idea is not to test a theory, but to find something that works. At all times, the client is very much in charge, and the science has to fit within political and organisational constraints.

The Behavioural Insights Team has made great play with trials, though it took about 18 months to fully establish them into its work programme. Trials were championed in the publication *Test, Learn, Adapt: Developing Public Policy with Randomised Controlled Trials* (Haynes *et al.*, 2012). The tax and fine reminder studies were done as trials (Behavioural Insights Team, 2012b), as was early work on charitable giving (Cabinet Office and Charitable Aid Foundation, 2013). Trials quickly became the signature of the unit and developed in scale and ambition over time, such as the trials of welfare-to-work at Loughton in Essex, which turned into a stepped-wedge trial throughout the whole county (see Halpern, 2015). The use of trials continued when the team was spun out of government, and this characterises its work on, for example, standards and attainment in education (see Behavioural Insights Team, 2015).

The demonstration work of the team probably encouraged a more general use of trials throughout UK government, where they became more routine. One example is a large-scale trial designed by the team and carried out by the Department for Business Innovation and Skills, testing whether receipt of online business support advice and a growth voucher would be good for businesses (Department for Business Innovation and Skills, 2015). Such use of trials has been encouraged by their rapid adoption in the US, in particular for education – work mirrored in the interventions commissioned by the Education Endowment Foundation, and other initiatives, such as the London Schools Excellence Fund, which has worked with the evaluation champion Project Oracle and its partnership organisation, The Social Innovation Partnership. At the same time, the government set up What Works research and evidence centres where robust forms of evaluation, such as randomised controlled trials, are privileged. It can be seen, then, that a variety of influences have promoted the use of trials in government and other agencies. The level of experience in running them has been growing strongly over time, making this kind of testing of programmes more feasible and giving a more scientific underpinning to evaluation. The use of trials is probably a more significant development than the use of behavioural sciences, and again, the expertise has come from economics while academic psychology does not feature very much and other disciplines, such as education and crime science, also contribute. In fact, there is a huge diversity in the kind of work being carried out.

Conclusion

This chapter started with the proposition that policy-making has been transformed by the use of ideas from the behavioural sciences, particularly from social psychology and behavioural economics. Taking the example of the UK, it is possible to observe such a development since the 2000s, even though psychologists have been employed in government in various contexts for many decades. The advance of the use of behavioural ideas has been gradual, with successive national governments taking an interest in the opportunities given by behavioural interventions to improve the delivery of policies. Governments like to hoover up new ideas and behavioural sciences made an easy entry based on simple applications, backed by a range of talented advocates within government as well as academics prepared to take to the airwaves. Behavioural techniques have been deployed across governments under the guidance of nudge units and other bodies; but it has been a pragmatic adoption focused on yielding instrumental benefits, often of a financial kind. Governments have tended to take a very instrumental approach to the science, and its adoption has fitted well with the interest of successive governments in the effective delivery of public services. More often than not, the modifications suggested by behavioural sciences, rather than being a straightforward application of behavioural science, are aided by the application of a lot of common sense (such as keeping things easy) and are influenced by proof of the concept through results from randomised controlled trials. At all times, the politicians and bureaucrats are in charge and own the whole process. Electoral interests, media pressure, party politics, ideology and other forms of expertise remain as the key determinants of the course of behavioural public policy and its implementation, as they do for all policies. The use of behavioural insights, as a simplified version of academic behavioural science, meshes well with the established interests of policy-makers. It has led to greater innovation in government and in the practice of public administration (John, 2014), but it cannot be said that there has been a major shift in policy as a result of behaviour change; rather, the change is in the way policy is delivered. Even the greater use of trials fits in with the UK government's long interest in the use of evidence and in its employment of social scientists throughout the departments of state and their agencies.

The chapter began with the debate about the emergence of the psychological state (Jones *et al.*, 2013a, 2013b) and how the term seems to imply a strong influence of behavioural ideas that reduce the autonomy of the citizen in the face of expertise and social scientific knowledge. Although the account presented in this chapter cannot, nor is it intended to, challenge this argument, it does show that the process takes a complex path. Such ideas enfold because of advocacy skills and the translation of social science arguments into the language and working assumptions of other policy-makers and public officials who work in the centre and in the complex array of government agencies that implement policy. It has been a gradual evolution, but one in which there have been flurries of interest. In this process, the work of academic psychology has not been prominent and a range of disciplines have contributed to the debate. In short, there is considerable agency in the diffusion of these ideas, and where politicians and bureaucrats make the case for behaviour change policy with their own preferences intact and existing values dominant.

Behaviour change policy is done in full public view, argued for in the media where members of the public can easily find out what has been decided on their behalf. The evolution of behaviour change fits well with a more general shift in focus in government to make public administration more citizen-centred when designing policies. It can work alongside measures that are designed to include citizens in debates about government decisions, so-called 'think' policies (John *et al.*, 2011). Psychological governance might, of course, work at a more subliminal level in terms of changes in working assumptions and practices over time. Policy-makers may internalise ideas from the behavioural sciences in their day-to-day work, consistent with the way in which ideas from the academy influence public policy (see John, 2013c), and maybe the influence is all the stronger because they have been so attractively and seductively presented. Nonetheless, the current state of play is more about attention to implementation and delivery in some selected areas. There is considerable enthusiasm, and political actors see it as being in their interest to follow these ideas; but standard operating procedures and the interplay of established interests remain as the defining characteristics of the public policy process in the UK and elsewhere.

Note

1 The author is a member of the panel. The comments and views expressed in this chapter are solely those of the author and should not be attributed to the Behavioural Insights Team.

References

Barber, M. (2008) *Instruction to Deliver: Fighting to Transform Britain's Public Services.* Methuen, London.

Becker, G. (1968) 'Crime and punishment: an economic approach'. *Journal of Political Economy*, 76(2): 169–217.

Behavioural Insights Team (2012a) Annual Update 2011–2012. Cabinet Office, London. Available at: https://www.gov.uk/government/uploads/system/uploads/atta chment_data/file/83719/Behavioural-Insights-Team-Annual-Update-2011-12_0.pdf

Behavourial Insights Team (2012b) Applying Behavioural Insights to Reduce Fraud, Error and Debt. Cabinet Office, London.

Behavourial Insights Team (2015) Behavioural Insights and the Somerset Challenge. Behavoural Insights Team, London.

Bell, C. (2013) 'Inside the coalition's controversial nudge unit', *The Telegraph*, 11 February.

Benartzi, S. and Thaler, R. H. (2004) 'Save more tomorrow: using behavioral economics to increase employee saving'. *Journal of Political Economy*, 112(1): 164–187.

Benjamin, A. (2013) 'David Halpern: we try to avoid legislation and ordering', *The Guardian*, 5 February.

Blume, T. and John, P. (2014) Using Nudges to Increase Council Tax Collection: Testing the Effects Through a Randomized Controlled Trial. Lambeth Borough Council, London.

Burch, M. and Holiday, I. (1995) *The British Cabinet System.* Harvester Wheatsheaf, Brighton.

Cabinet Office and Charitable Aid Foundation (2013) Applying Behavioural Insights to Charitable Giving. Cabinet Office, London. Available at: https://www.gov.uk/govern

ment/uploads/system/uploads/attachment_data/file/203286/BIT_Charitable_Giving_
Paper.pdf (accessed 2 August 2015).

Department for Business Innovation and Skills (2015) Growth Vouchers Programme:
Phase One Qualitative Assessment Report. BIS Research Paper No. 220. Department
for Business Innovation and Skills, London.

Dolan, P., Hallsworth, M., Halpern, D., King, D. and Vlaev, I. (2010) MINDSPACE:
Influencing Behaviour for Public Policy. Institute for Government/Cabinet Office,
London.

Dubner, S. J. and Levitt, S. (2005) *Freakonomics: A Rogue Economist Explores the
Hidden Side of Everything.* Penguin, London.

Duflo, E., Glennerster, R. and Kremer, M. (2006) Using Randomization in Development Economics Research: A Toolkit. Technical Working Paper No. 333. National
Bureau of Economic Research, Cambridge, MA.

Dunlop, C. (2013) 'The costs of mining evidential gold: randomized control field trials
and policy-relevant knowledge'. Paper presented to the 2013 Policy and Politics
conference, Transforming Policy and Politics – The Future of the State in the 21st
Century, Bristol, 17–18 September.

Edgerton, D. (2009) 'The "Haldane Principle" and other invented traditions in science
policy' [online]. *History and Policy*, 2 July. Available at: www.historyandpolicy.
org/policy-papers/papers/the-haldane-principle-and-other-invented-traditions-in-
science-policy

Eisenstadt, N. (2011) *Providing a Sure Start.* Policy Press, Bristol.

Fisher, R. A. (1935) *The Design of Experiments.* Oliver and Boyde, Edinburgh.

Gueron, J. M. and Rolston, H. (2013) *Fighting for Reliable Evidence.* Russell Sage
Foundation, New York.

Hallsworth, M., List, J. A., Metcalfe, R. D. and Vlaev, I. (2014) The Behavioralist as
Tax Collector: Using Natural Field Experiments to Enhance Tax Compliance.
Working Paper No. w20007. National Bureau of Economic Research, Cambridge,
MA. Available at: http://ssrn.com/abstract=2418122

Halpern, D. (2015) *Inside the Nudge Unit.* W. H. Allen, London.

Halpern, D. and Bates, C. (2004) *Personal Responsibility and Changing Behaviour.*
Prime Minister's Strategy Unit, London.

Haskins, R. and Margolis, G. (2015) *Show Me The Evidence.* The Brookings Institution,
Washington, DC.

Haynes, L., Service, O., Goldacre, B. and Torgerson, D. (2012) *Test, Learn, Adapt:
Developing Public Policy with Randomised Controlled Trials.* Cabinet Office, London.

Haynes, L., Green, D. P., Gallagher, R., John, P. and Torgerson, D. (2013) 'Collection
of delinquent fines: a randomized trial to assess the effectiveness of alternative
messages'. *Journal of Policy Analysis and Management*, 32(4): 718–730.

H. M. Treasury (2011) *The Magenta Book: Guidance for Evaluation.* HMSO, London.

House of Lords (2011) Science and Technology Sub-Committee, 2nd Report of Session
2010–12, Behaviour Change. The Stationery Office, London.

Independent Scientific Group on Cattle TB (2007) Bovine TB: The Scientific Evidence.
A Science Base for a Sustainable Policy to Control TB in Cattle. An Epidemiological
Investigation into Bovine Tuberculosis. Final Report of the Independent Scientific
Group on Cattle TB Presented to the Secretary of State for Environment, Food and
Rural Affairs, The Rt Hon David Miliband MP. Defra, London.

John, P. (2013a) 'All tools are informational now: how information and persuasion
define the tools of government'. *Policy and Politics*, 41(4): 605–620.

John, P. (2013b) 'My nudge tour of Australia' [online]. Available at: http://publicpolicy. southampton.ac.uk/nudge-tour/

John, P. (2013c) 'Political science, impact and evidence'. *Political Studies Review*, 11(2): 168–173.

John, P. (2014) 'Policy entrepreneurship in British government: the Behavioural Insights Team and the use of RCTs'. *Public Policy and Administration*, 29(3): 257–267.

John, P. (forthcoming) *Field Trials in Political Science and Public Policy*. Routledge, London.

John, P. and Richardson, L. (2012) *Nudging Citizens Toward Localism*. British Academy, London.

John, P. and Blume, T. (2015) Using Nudges to Increase Online Blue Badge Renewal – Testing the Effects through a Randomised Controlled Trial for Essex County Council. Essex County Council, Chelmsford.

John, P., Cotterill, S., Moseley, A., Richardson, L., Smith, G., Stoker, G. and Wales, C. (2011) *Nudge Nudge, Think Think: Experimenting with Ways to Change Civic Behaviour*. Bloomsbury Academic, London.

Jones, R., Pykett, J. and Whitehead, M. (2013a) *Changing Behaviours: On the Rise of the Psychological State*. Edward Elgar, Cheltenham.

Jones, R., Pykett, J. and Whitehead, M. (2013b) 'Psychological governance and behaviour change'. *Policy and Politics*, 41(2): 159–182.

Kahneman, D. (1973) *Attention and Effort*, Prentice-Hall, Englewood Cliffs, NJ.

Kahneman, D. (2011) *Thinking, Fast and Slow*. Penguin, London.

Kahneman, D. and Tversky, A. (1979) 'Prospect theory: an analysis of decision under risk'. *Econometrica*, 47(2): 263–291.

Kahneman, D., Slovic, P. and Tversky, A. (1982) *Judgment Under Uncertainty: Heuristics and Biases*. Cambridge University Press, Cambridge.

King, D. (1998) 'The politics of social research: institutionalizing public funding regimes in the United States and Britain'. *British Journal of Political Science*, 28(3): 415–444.

Kingdon, J. W. (1984) *Agendas, Alternatives, and Public Policies*. Little, Brown, Boston.

Leggett, W. (2014) 'The politics of behaviour change: nudge, neoliberalism and the state'. *Policy and Politics*, 42(1): 3–19.

Majone, G. (1989) *Evidence, Argument, and Persuasion in the Policy Process*. Yale University Press, New Haven, CT.

Marteau, T. M., Ogilvie, D., Roland, M., Suhrcke, M. and Kelly, M. P. (2011) 'Judging nudging: can nudging improve population health?' *British Medical Journal*, 342: 263–265.

Newell, B., Lagnado, D. and Shanks, D. (2007) *Straight Choices*. Psychology Press, Hove, UK.

Oliver, A. (2013a) *Behavioural Public Policy*. Cambridge University Press, Cambridge.

Oliver, A. (2013b) 'From nudging to budging: using behavioural economics to inform public sector policy'. *Journal of Social Policy*, 42(4): 685–700.

Park, A. and Sabourian, H. (2011) 'Herding and contrarian behavior in financial markets'. *Econometrica*, 79(4): 973–1026.

Service, O., Hallsworth, M., Halpern, D., Algate, F., Gallagher, R., Nguyen, S., Ruda, S., Sanders, M., with Pelenur, M., Gyani, A., Harper, H., Reinhard, J. and Kirkman, E. (2014) EAST: Four Simple Ways to Apply Behavioural Insights. BIT/Cabinet Office, London.

Shafir, E. (Ed.) (2013) *The Behavioral Foundations of Public Policy.* Princeton University Press, Princeton, NJ.

Shah, H. and Dawney, E. (2005) *Behavioural Economics: Seven Principles for Policy-makers.* New Economics Foundation, London.

Strassheim, H. and Korinek, R. (2016) 'Cultivating "nudge": behavioural governance in the UK', in J.-P. Voß and R. Freeman (Eds) *Knowing Governance: The Epistemic Construction of Political Order.* Macmillan, Basingstoke, pp. 107–126.

'Student' (1931) 'The Lanarkshire milk experiment'. *Biometrika*, 23(3/4): 398–406.

Thaler, R. H. and Sustein, C. R. (2008) *Nudge: Improving Decisions about Health, Wealth and Happiness.* Yale University Press, New Haven, CT.

Weiss, C. H. (1972) 'The politicization of research', in C. H. Weiss (Ed.) *Evaluating Action Programs: Readings in Social Action and Education.* Prentice Hall, Boston, pp. 327–338.

White, M. and Lakey, J. (1992) *The Restart Effect: Evaluation of a Labour Market Programme for Unemployed People.* Policy Studies Institute, London.

Index

Milton Keynes UK
Ingram Content Group UK Ltd.
UKHW031151141024
449569UK00024B/884